THE BOOK II

More Engineering Guidance for the Digital Transition

An NVISION® Guide

ISBN 0-9640361-7-7

Printed in the United States of America.

Contents

Acknowledgments	v
Preface	vii
The Digital Television Environment	1
Signal Distribution	15
Multi-Channel Audio	29
Routing Systems	59
Control Systems	75
Embedded Audio	99
Fiber Optic Signal Distribution	109
Bringing It All Together	131

Acknowledgements

This book has been written from the consolidated experiences of design and systems engineers involved in the continual development of our industry. Our thanks to all those who have unwittingly contributed to our knowledge base as well as those who either volunteered or were tricked into working on this project.

Our particular thanks to Steve Lyman of Dolby Laboratories for his excellent explanation of Dolby multi-channel audio compression in chapter 3. Thanks also to Geoff Ward and Paula Gleicher for writing chapters 5 and 7, Rob Bowdish for his diagrams, and the rest of NVISION for helping proof and edit what we hope will be a useful addition to your engineering library. Last, but by no means least, thanks to Jill Mahanna for her speedy book design and grammatical expertise.

Preface

In 1996 we published *THE BOOK: An Engineers Guide to the Digital Transition.* This publication dealt mainly with digital audio for video, as digital audio signals were the least understood and therefore the most difficult to manage. This situation has changed a little as the use of digital audio has become more commonplace, new DTV/DVB systems actually add more stringent audio requirements than previously faced.

Following numerous requests from readers of *THE BOOK* we have published this engineering guide to cover the new topics that have arisen, particularly in light of a worldwide requirement to digitize television broadcasting. Digital television systems require us all to learn new technologies and implementation techniques, the objective of this book is to help provide some insight into the difficulties faced when creating and managing an all digital television system.

Naturally, you will find that these pages make frequent reference to NVISION products and their application.

We trust that you will find this book helpful. We would be grateful for any comment regarding its content or suggestions for subject matter that you would like discussed in future publications.

Nigel Spratling
Vice President, Marketing
Author/Editor

NigelSpratling@NVISION1.com

Chapter 1
The Digital Television Environment

Digital technology has been employed for more than two decades in television production and post-production systems. Digital products have been designed to operate with the same picture and sound format as their analog counterparts, but provide better signal performance (particularly after multiple recording generations) as well as powerful manipulation and editing capabilities that are impossible in an analog domain.

Today, digital technology provides program producers with almost limitless creative tools and techniques. It allows audio and video quality to be controlled through every part of the creative process, ensuring that the final product meets the requirements of its producer. However, once this material is delivered to the distribution channels (terrestrial broadcasters, satellite, cable services or videotape), it is converted to its analog equivalent and quality control is at the mercy of the delivery system used. Terrestrial broadcasts are plagued by poor signal reception properties, cable systems often suffer from cross-modulation problems and analog consumer receivers deliver less than perfect results to the viewer, regardless of input signal quality.

After years of debate, many countries have elected to replace their analog delivery systems with new digital services. These new digital delivery services provide a level of flexibility that could not have been considered in an analog environment. Such flexibility includes choice of picture quality and aspect ratio, audio channel choices, and data services such as electronic program guides and interactive responses. All of the new services include the provision to deliver High Definition Television (HDTV), Standard Definition Television (SDTV) and in some cases, Enhanced SDTV (EDTV) programming, with choices for line and field rates, picture aspect ratio and audio format.

Implementing systems to provide suitable output formats for the intended delivery medium presents a number of challenges:

1. New digital services may need to accommodate multiple sound and picture formats.

2. Different compression techniques will need to coexist within a single system.

3. Compression systems may generate complex signal latency problems.

4. Surround sound audio will bring new signal management constraints.

5. Analog services will continue to exist for many years (probably ten or more).

The purpose of this book is to help identify solutions to these challenges, while highlighting the potential pitfalls that may face those of us who have the task of designing, constructing, or operating these systems that will provide new digital television services.

Picture and Sound Formats

Until now, the world has been used to two basic picture formats, 525/59.94 and 625/50, both with 2:1 interlace. Compatibility with existing systems, technical issues, and politics are factors that have determined the next generation of formats. Technically, we would all like to see a single standard adopted worldwide. Our lives would be much simpler and we'd need to know less. However, this is not to be; standards are much more complex and have probably provided us with ongoing job security.

Two bodies, the ATSC (Advanced Television Systems Committee) in the USA and the European DVB (Digital Video Broadcasting) project, have each developed recommendations for picture, sound and signal formats. These have been adopted and ratified as standards by the various government agencies that manage national communications (e.g. the Federal Communications Commission (FCC) in the USA).

DVB formats provide a simple choice: one HDTV format, one EDTV format, and two SDTV formats. See figure 1-1. The ATSC system offers a large choice of picture formats that can be delivered to the home and ATSC receivers must be able to display them all.

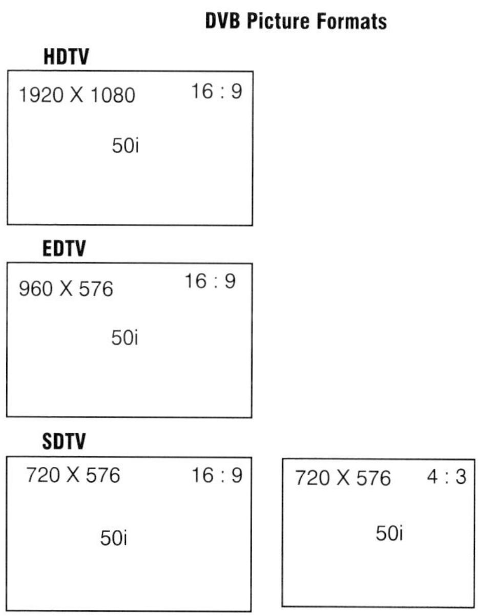

Fig 1-1. DVB Picture Formats (i = interlaced)

The choice and number of formats within the ATSC system can be quite confusing. When researching this book, we found frequent reference to "table 3" of the ATSC document A/53, and to the 18 picture formats to which it refers. Figure 1-2 shows how these 18 formats are derived.

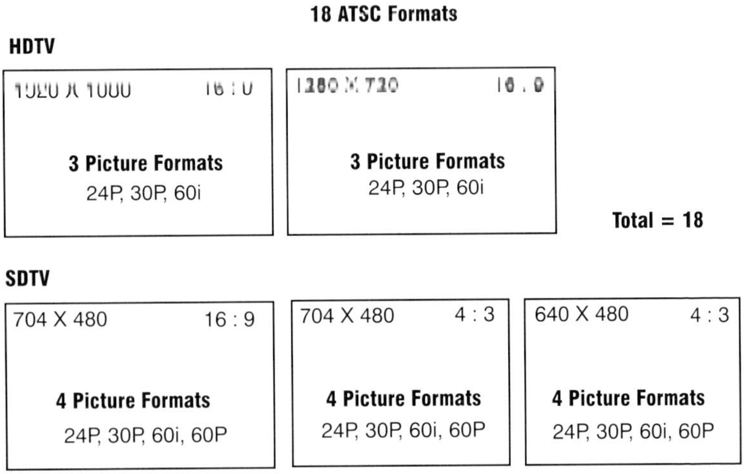

Fig 1-2. ATSC (Advanced Television Systems Committee) Formats

Unfortunately, the ATSC table is misleading because it assumes that integer frame rates are the same as non-integer rates (i.e. 30/29.97Hz). This is true from the viewer's perspective, but the frame rates are distinctly different technically.

Ideally, all frame rates should be separately identified, which means that the actual number of picture formats is 36. See figure 1-3.

So the real number of picture formats that can be transmitted to the home in an ATSC environment is 36? Yes, but not all of them can be transported within existing digital transport standards.

36 ATSC Formats

HDTV

1920 X 1080 16 : 9	1280 X 720 16 : 9
6 Picture Formats 24P, 30P, 60i, 23.976P, 29.97P, 59.94i	**6 Picture Formats** 24P, 30P, 60i, 23.976P, 29.97P, 59.94i

Total = 36

SDTV

704 x 480 16 : 9	704 X 480 4 : 3	640 X 480 4 : 3
8 Picture Formats 24P, 30P, 60i, 23.976P, 29.97P, 59.94i, 59.94P, 60P	**8 Picture Formats** 30P, 60i, 23.976P, 24P, 29.97P, 59.94i, 59.94P, 60P	**8 Picture Formats** 30P, 60i, 23.976P, 24P, 29.97P, 59.94i, 59.94P, 60P

Fig 1-3. All possible picture formats

Signal Transport Formats

Can all of these possible picture formats be carried by the agreed-upon digital transport standards? The answer is no, not at standard sample rates. Refer to the following example:

> A standard definition, 525/29.97, 4:2:2 component signal has a sample rate of 13.5MHz. Therefore;
>
> 13.5MHz / 29.97 = 450450 samples per frame
>
> 450450 / 525 = 858 samples per line
>
> However, the same signal with a 30Hz frame rate works out like this;
>
> 13.5MHz / 30 = 450000 samples per frame
>
> 450000 / 525 = *857.1428571429* samples per line

This does not work. In order to distribute standard definition signals via a normal serial digital transport stream (SMPTE 259M/ITU 601), only frame rates of 29.97 may be used. In fact all 525 line signals with a 13.5MHz sample rate must use frame rates based on 30/1.001 (29.97). Further investigation reveals that the only 640 x 480 rates that could be transported would have to have a 29.97 frame rate *and* be re-mapped into SMPTE 259M.

Figure 1-4 shows the ATSC standard definition formats that can be transported via existing digital interfaces:

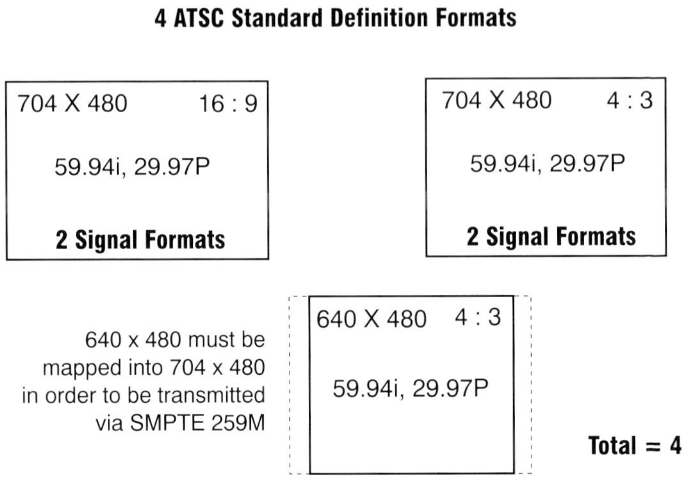

Figure 1-4. 4 ATSC Standard Definition Formats

High definition signals can be derived from two sample rates: 74.25MHz for integer frame rates and 74.25MHz/ 1.001 for 525 line compatibility. This means that, unlike SDI formats that share a common data rate (270Mbits), HD-SDI signals have two possible rates. It is important to remember that the generic reference to 1.5 Gbits for HD transport does not tell you what the actual data rate is. Figure 1-5 shows the calculation for the HD-SDI data rates. Table 1-1 provides a complete list of all of the formats that are transportable over standard interfaces. Table 1-2 shows all of the current signal transport methods that may be implemented today.

Interestingly, many different picture formats can be carried via a common serial interface. When researching this book, we discovered that relatively few people understand how these picture formats can share the same sample rate yet be so different. Figure 1-6 provides a graphical representation of exactly how these various formats coexist.

Data rate @ 30 frames =
Sample rate x 2 (1 luma sample + 2 chroma samples @ half the luma rate) x 10 (with a 10 bit sample size)
= 1.485 Gbits
Data rate @ 29.97 frames =
Sample rate (74.25Mhz/1.001) x 2 x 10
= 1.48316483516 Gbits (recurring decimal)

Figure 1-5. HD-SDI serial data rates

Figure 1-6. How active picture formats coexist

Table 1-1. Transportable Picture Formats

Picture Formats					Transport Interface
Active Vertival Lines	Active Pixels per Line	Picture Aspect Ratio	Picture Scanning Method *		Baseband Serial Data Rate
1080	1920	16:9	60Hz fields	I	SMPTE 292M
1080	1920	16:9	59.94Hz fields	I	**1.5 Gbits
1080	1920	16:9	50Hz fields	I	
1080	1920	16:9	30Hz frames	P	
1080	1920	16:9	29.97Hz frames	P	
1080	1920	16:9	24Hz frames	P	
1080	1920	16:9	23.97Hz frames	P	
720	1280	16:9	60Hz frames	P	
720	1280	16:9	59.94Hz frames	P	
720	1280	16:9	30Hz frames	P	
720	1280	16:9	29.97Hz frames	P	
720	1280	16:9	24Hz frames	P	
720	1280	16:9	23.97Hz frames	P	
576	960	16:9	50Hz fields	I	SMPTE 259M
					360Mbits
576	720	16:9	50Hz fields	I	SMPTE 259M
576	720	4:3	50Hz fields	I	270 Mbits
480	704	16:9	59.94Hz frames	P	SMPTE RP175
480	704	4:3	59.94Hz frames	P	***540 Mbits
480	704	16:9	59.94Hz fields	I	SMPTE 259M
480	704	4:3	59.94Hz fields	I	270 Mbits
480	704	16:9	29.97Hz frames	P	
480	704	4:3	29.97Hz frames	P	

** I = interlace, P = progressive*
*** 1.5 Gbits is normally quoted as the generic data rate for HDTV signals.*
**** At this time 540 Mbit transmission requires two 270 Mbit serial links (SMPTE Recommended Practice 175). However, it is likely that a 540 Mbit standard will be published in the future.*

Table 1-2. Current Signal Transport Methods

Baseband Signal Type	Frequencies/ Serial Data Rate	Sample Rate	STD
Analog NTSC/PAL (and derivatives)	4.2–5.5 MHz	N/A	
SDI Composite NTSC	143 Mbits	14.318 MHz	259M/A
SDI Composite PAL	177 Mbit	17.734 MHz	259M/B
SDI Component 525/625 4:3 (16:9 w/stretched pixels)	270 Mbit	13.5 MHz Y 6.75 MHz Cr,Cb	259M/C
SDI Component 525/625 16:9	360 Mbit	18 MHz Y 9MHz Cr,Cb	259M/D
HD-SDI @ 23.976/29.97/ 59.94 frame/field rates	1.483516-Gbits	74.1758- MHz Y 37.0879- Cr,Cb	292M
HD-SDI @ 24/25/30/ 50/60 frame/field rates	1.485 Gbits	74.25 MHz Y 37.125 Cr,Cb	292M
Analog audio (1 channel)	0 –20kHz	N/A	
AES/EBU audio (2 channels)	3.072 Mbits	48 kHz	AES3

Program production will require equipment that is format specific or multi-format, as most picture formats are not directly compatible or interchangeable. Any system design needs to accommodate camera, storage and editing equipment that is specific to the picture and sound formats supported, even though the connectivity requirements can be common.

So far we have only mentioned video, but the audio portion of the digital television environment presents some new and difficult issues. In both the ATSC and DVB systems, the audio capacity is 5.1 channels and can include mono mix, stereo, pro logic, multiple language, and full surround sound.

Signal compression systems have been developed and standardized for delivery to the consumer (Dolby Digital (AC3) and MPEG2), but the transport, storage and editing of baseband surround sound signals requires a minimum of six channels (three AES signals). Currently, standard DVTRs have no more than a four channel capacity (with the exception of some highly specialized HDTV recorders). Furthermore, if video storage devices exist (i.e. digital disk recorders) that can accommodate the channel requirement, maintaining accurate phase alignment presents some complex difficulties.

Interfaces are defined and hardware is under develpment for managing multi-channel audio as a single data stream at baseband or compressed at a 'mezzanine' level. These techniques will be discussed in detail in chapter 3.

Obviously, the choice of signal formats to be managed is determined by each application. A single standard may suffice for certain applications, while others may need to handle a wide variety.

Production / Post Production

Any facility providing production or post-production services is faced with a difficult dilemma. In order to provide the services required, several formats may need to be managed. Ideally, a common production format from which all other formats could be output would present a perfect solution. Today, the only common format is film (it may vary in size but the Tele-cine operator can handle that), but editing celluloid and creating optical effects is uneconomic, undesirable and in many cases impossible.

Currently, dual-format post-production (525/625) is commonplace for international distribution. It is not uncommon for film to be transferred to video twice (once for each format) so that video standards conversion is avoided. Achieving acceptable conversions at a high quality is a very specialized task and many producers still prefer the look of a program that is transferred from film in the final video format. This requires that most programs are transferred and edited twice, therefore the costs are significantly higher than programs produced for local distribution only. With the rapid expansion of world communication systems, the requirement for international program distribution is increasing. The addition of new picture formats could potentially add significant costs to the post production portion of program generation, as well as require the design of very complex technical systems to accommodate all of the required standards.

Discussions within the post community have focused on generating a single electronic 'film' format that could be used during all production and post-production stages and then be converted to the desired delivery format for each distribution channel after the product is complete. The ideal candidate for this single format is 1080P/24; it is the direct electronic equivalent of film and is easily converted to all of the possible frame rates, without the introduction of undesirable motion artifacts. Simple line interpolation can be used to derive any of the possible line rates. Figure 1-7 shows how 1080P/24 (23.97 for 1/1001 formats) can be converted to 30/29.97 based formats by a process of repeating fields or frames. This process does generate a perceivable motion "judder" at certain object speeds and direction. (Judder is the result of film frames being represented alternately by two and three frames/ fields, as shown in figure 1-7.) But it is generally acceptable and has been a standard practice for all 24 frame film to NTSC transfers. To convert to a 25-frame format requires a 4.1% speed change, which affects running time and adds a pitch shift to the audio content.

At the time of writing this book, several products have been shown and demonstrated for 1080P/24 production and editing. Post production systems will be discussed further in chapter 2.

Broadcasting Networks

In most broadcasting systems, only one or two picture formats will be output for delivery to the viewing audience. So, system requirements will be simpler than those of the post production facility. However, most broadcasting to date has been single standard, and so we are considerably increasing the complexity with the addition of DTV/DVB.

Managing HD and SD within the same system could be as simple as including up and down converters at the facility input and output, or as complex as separate distribution and production layers for each format.

Naturally, the choice of formats and operational methods is largely up to the broadcasting organization, as the viewer will be able to receive the broadcast regardless of chosen format. Implementing a system designed to manage the processing and distribution of the desired signals could be very complex and costly and serious consideration needs to be given to the design. In the next chapter we will discuss some of the design and implementation issues.

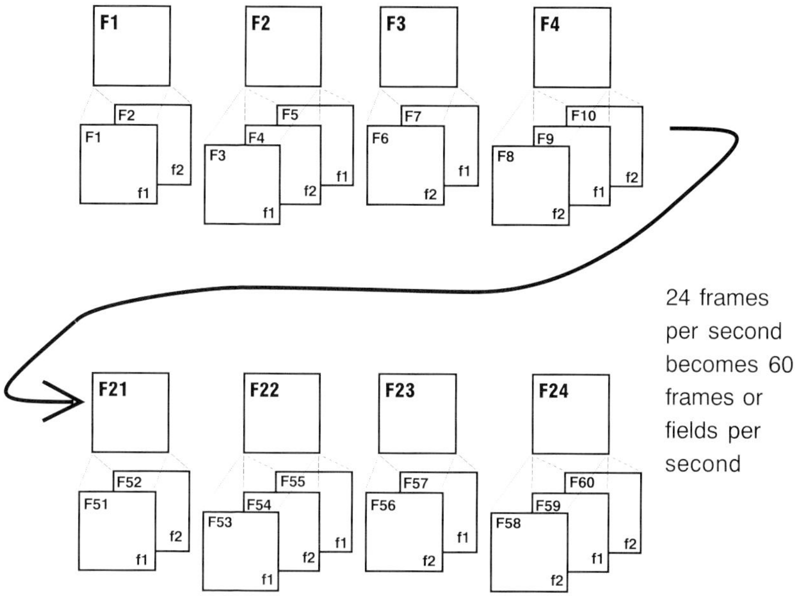

Figure 1-7. A typical conversion sequence from 24P to 60i and 60P using 3:2 pulldown (F = frame, f = field)

Chapter 2
Signal Distribution

SDI Signals

SDI, the Serial Digital Interface as described by SMPTE and ITU standards, has been with us for a number of years and there are many mature products available that provide functional interfaces. The SDI standards allow that data rates of 143, 177, 270 and 360 Mbits share this common interface. Distribution amplifiers, routing switches, patch bays and coaxial cables that are supplied for SDI should handle all of these rates; however, this is not always true. SDI standards and interface technology have progressed over the years and early product designs were often specific to a data rate. Most currently available product will adequately pass 143 through 270 Mbit data. If you need to accommodate 360 Mbit signals, be cautious, as not all cable equalizers and reclocking devices work at this rate. Naturally, maximum usable cable length is determined by the data rate, quality of cable, and receiver type. Table 2-1 shows typical transmission length for cable type and data rate.

The NVISION 4000 Series distribution amplifiers are designed using the latest parts available and are guaranteed to function at all SDI data rates. The SD4110 is a simple eight-output fanout DA with auto cable EQ. The SD4111 adds reclocking, for installations where longer cable lengths and higher jitter tolerance are required.

Table 2-1. Maximum Cable Lengths for Transmitting SDI (Data Rate/Cable Type)

Data Rate	143 Mb/s		177Mb/s		270 Mb/s		360 Mb/s		540 Mb/s	
Belden Part #	Ft.	M	Ft.	M	Ft.	M	Ft.	M	Ft.	M
1865A	730	222	658	201	540	165	467	142	380	116
8279	818	249	725	221	574	175	493	150	385	117
1855A	915	279	818	249	659	201	572	174	477	146
9209	931	284	834	254	675	206	587	179	482	147
9209A	931	284	834	254	675	206	587	179	482	147
1505A	1286	392	1180	360	1000	305	868	265	711	217
1506A	1227	374	1079	329	844	257	732	223	593	181
9231	1286	392	1138	347	900	274	768	234	659	201
9141	1286	392	1138	347	900	274	768	234	659	201
8281	1286	392	1138	347	900	274	775	236	625	191
8281B	1286	392	1138	347	900	274	768	234	614	187
8281F	1102	336	970	296	761	232	654	199	529	162
88281	1174	358	1037	316	818	249	689	210	540	165
1694A	1588	484	1443	440	1200	366	1051	320	870	265
1695A	1500	457	1359	414	1125	343	974	297	794	242
7731A	2477	755	2236	681	1837	560	1587	484	1291	394

Table Courtesy Belden Cable; other manufacturers' equivalents typically will have similar specifications

It is often necessary to transport SDI signals over greater distances than practical on a single coax cable. This can be achieved by placing reclocking DAs in the path as signal repeaters (figure 2-1). However, using repeaters is not always practical so an excellent alternative is to transport the signal via fiber optic link. In the past, such links have been expensive to install and have not necessarily handled all possible signal content flawlessly. NVISION has designed a new generation of fiber optic conversion devices that combine correct technical performance with affordability. Information on these products, their application and the use of fiber in general is discussed in chapter 7.

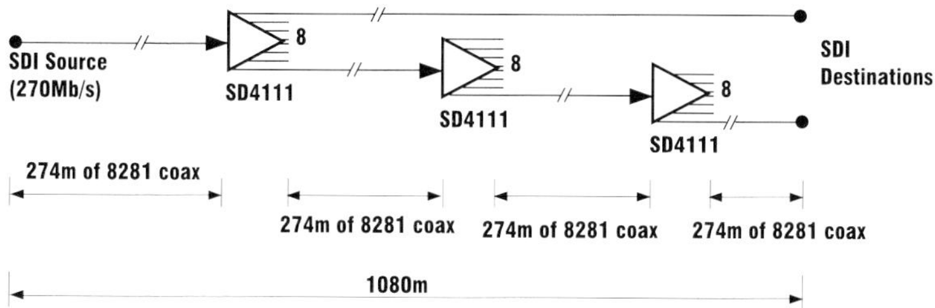

Figure 2-1. Using the SD4111 reclocking DAs as repeaters

Modern routing switches designed specifically for SDI should pass all data rates. However, it's an excellent idea to check data rates before a purchase. If you are constructing a system with long cable lengths, be aware that many routers do not offer reclocking and may require the addition of reclocking DAs. If you need to accommodate a mixture of data rates and field/frame rates, your choices are limited. Chapter 4 is dedicated to routing switches and managing multiple signal types.

Jitter, Pathological Content and Polarity

SDI signals are generally immune to the causes of disruption that damage their analog predecessors. Minor level fluctuations, common mode interference, signal polarity and induced hum have little or no effect on the receiveability of SDI signals. However, jitter induced by transmission lines or clock instability can cause errors at the receiver that vary from occasional bit errors (pixel drop-outs or sparkles) to complete reception failure.

Care should be taken to avoid transmission line jitter by ensuring that good quality coaxial cables are selected, that lengths are kept within manufacturers' recommendations and that patch bays and terminations are of the correct impedance. There are still some patch bays and terminators sold for video applications that have a nominal impedance of 50 ohms. If used they may induce jitter that is only apparent in certain paths, due to the accumulation of transmission line and clock jitter. It may be very difficult to identify the source of the problem.

When the serial digital interface was developed, it was necessary to scramble the data to remove the possibility of recurring data patterns of 1s or 0s that could potentially cause a temporary shift in the DC level. If a significant DC shift occurred it could reduce the receiver's ability to recover the signal data. To overcome this, a nine-bit polynomial filter was included in the serializer design. At that time a video sample was eight bits deep, and so the signal could be effectively scrambled to remove any repetitive patterns. See figure 2-2.

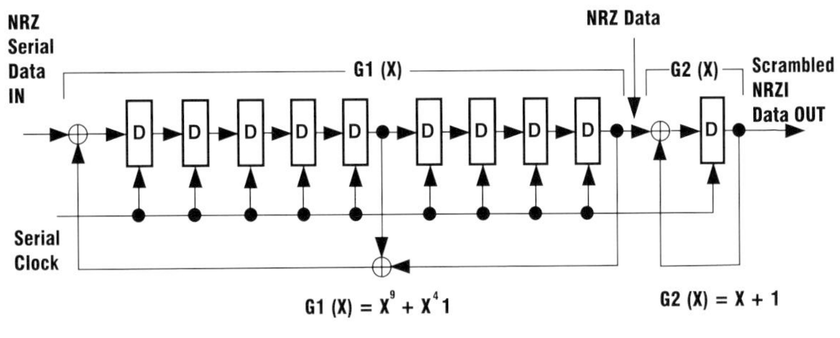

Figure 2-2. A polynomial scrambler

When the sample depth was changed to 10 bits, two specific combinations of luma and chroma values caused the nine-bit scrambler to generate a string of a 1 and 19 0s, or 2 2s and 18 0s. These two are referred to as pathological values. As these samples would theoretically need to be repeated consistently to cause a significant DC shift, the SDI standards were left unchanged. In most situations, repeated pathological samples should not result in errors, however, a poorly designed receiver may translate these signals into visible bit errors or 'sparkles'. (EQ Stress and PLL Stress are pathological test signals. Check Field is a split field test signal with half a field of EQ stress followed by half a field of PLL stress.)

The SDI data stream is NRZI (Non Return to Zero Inverting) coded, shown in figure 2-3. NRZI coding provides circuit designers with the ability to use both positive and negative outputs of the differential amplifiers in circuit designs, which significantly reduces component count and therefore product costs. It is not normal practice to identify which paths within a design invert a signal.

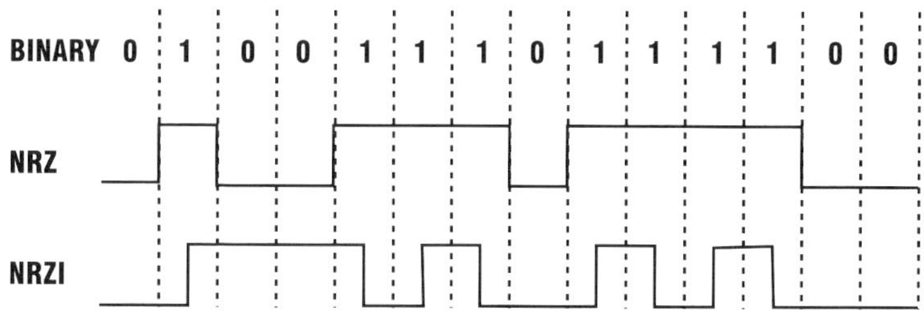

Figure 2-3. NRZ and NRZI coding schemes

ASI and SDTI

Standards have been defined for the transport of multiplexed packets of compressed video data, DVB-ASI and SDTI. These schemes allow multiple video signals to be transported over a single SDI type connection at 270 or 360 Mbits. SDTI follows the same signal conventions as SDI and can be transported by standard SDI products.

ASI is very similar in that it too can be transported via standard SDI distribution methods; however, ASI is NRZ (Non Return to Zero) coded and cannot be inverted. (Figure 2-3.) This presents system designers handling ASI data with somewhat of a problem, as almost all SDI products utilize differential amplifiers. Often the system designer cannot tell, without delving into schematics, which paths invert. This means that even the humble DA with 6, 8 or 10 outputs is reduced to 3, 4 or 5 useable outputs and at that, it's 50/50 as to which ones they are.

All NVISION SDI products are designed with the inverted paths clearly identified where inversion exists, and with special attention given to ensuring that there is no inversion in primary signal paths. This makes system design for the inclusion of ASI signals straightforward.

HD-SDI and HD-SDTI

The HD-SDI and HD-SDTI are basically much higher bandwidth versions of SDI and SDTI, The bit rates are much faster (1.5 Gbits) and therefore need to be handled with much greater care than the standard definition counterpart.

Clock stability, transmission line impedance and return losses become far more critical at HD-SDI data rates. Cable type and length, the quality of physical termination (how accurately the BNC is attached), and the choice of patch bays and terminators are all critical factors in creating an error-free system. At 1.5 Gbits, small cable stubs and short unshielded conductors can become large reflectors and a source of error-inducing jitter. Poor return loss in a receiver design can reflect back into the transmitting device, generating show-stopping errors.

Table 2-2 shows recommended cable lengths for transporting 1.5 Gbits via a selection of coax types.

Table 2-2. Maximum Cable Lengths for Transmitting 1.5Gbits

Data Rate	1.5 Gb/s	
Belden Part No.	Ft.	M
1865A	150	46
8279	153	47
1855A	193	59
9209	182	55
9209A	177	54
1505A	272	83
1506A	241	73
9231	233	71
9141	233	71
8281	238	73
8281B	225	69
8281F	212	65
88281	181	55
1694A	335	102
1695A	282	86
7731A	494	151

Table Courtesy Belden Cable; other manufacturers' equivalents typically will have similar specifications

Jitter is the most common cause of bit errors in the HD-SDI system, and not all products are equal in their ability to minimize or overcome these errors. SMPTE 292M specifies that output jitter should not exceed 134 ps. Our tests of various HD-SDI origination devices have revealed that many currently available products generate a level of jitter that is not always within SMPTE 292M specifications. If these signals accumulate small amounts of additional jitter in the transport path, bit errors will likely result.

Figure 2-4 shows the 'eye' pattern measurements taken from various HD-SDI sources. Note that the jitter levels are in excess of SMPTE specifications.

We have discovered that most output designs that include dual or multiple outputs can exhibit increased jitter levels

(a)

(b)

(c)

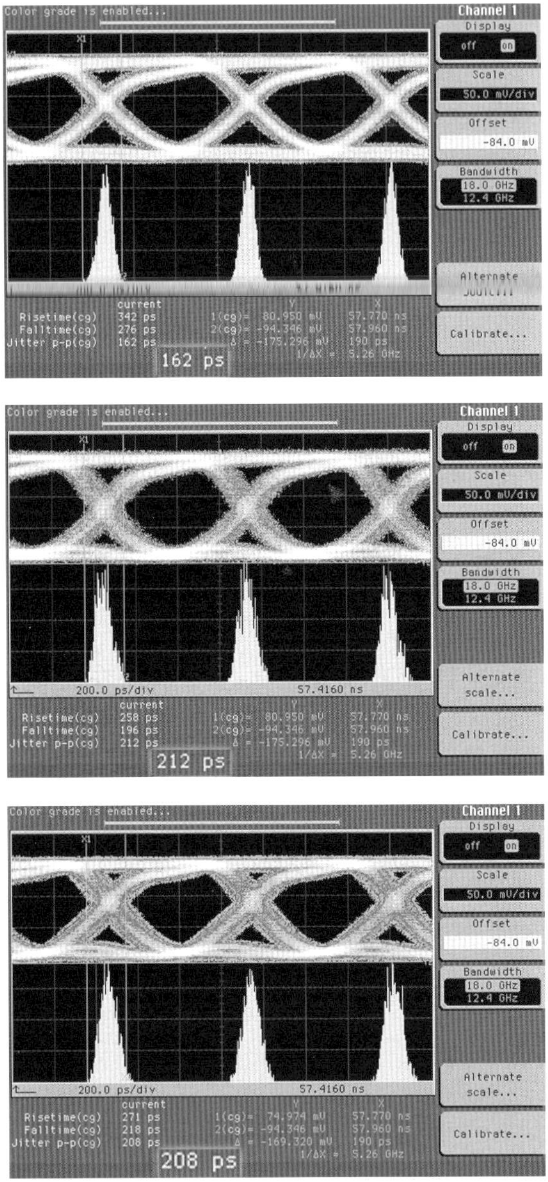

Figure 2-4. Measured output jitter from (a) HD VTR (b) HD camera (c) effects generator

(a)

(b)

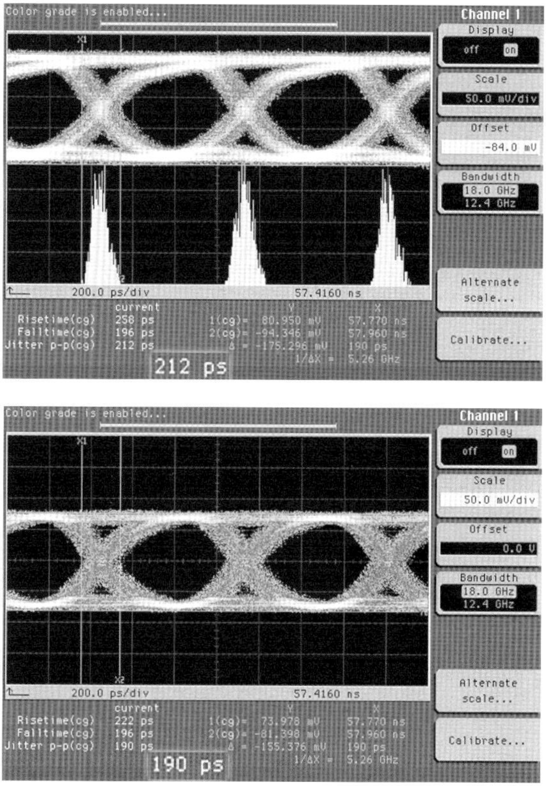

Figure 2-5. Signal through a well-known manufacturer's reclocking DA (a) Before reclocking (b) After reclocking

if unused outputs are not terminated. These unused outputs can easily become a source of reflections that will feed back into the output amplifier unless it is correctly loaded.

Reclocking requires the use of a phase locked loop (PLL) with an accurate oscillator. At 1.5 Gbits, designing an oscillator with a sufficiently low jitter level is quite difficult (there is always some level of clock instability at any frequency). Minimizing clock jitter is the key to designing an effective reclocker. Again, not all designs are equal. Figure 2-5 shows a before and after measurement, through a well-

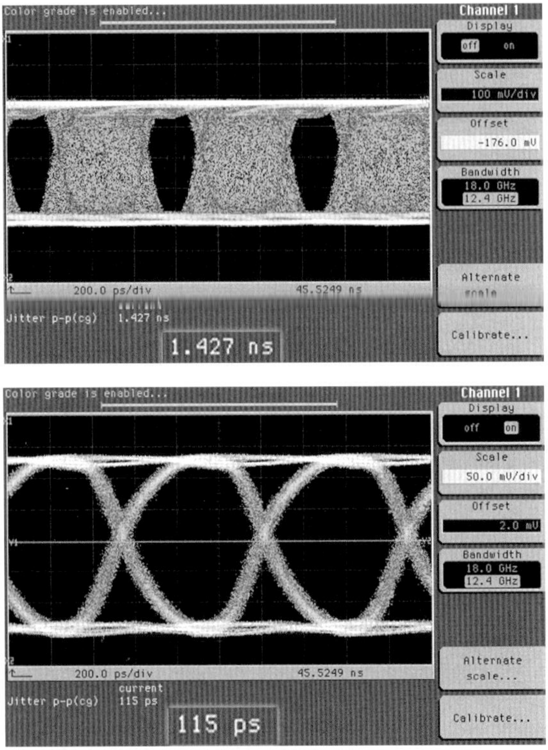

(a)

(b)

Figure 2-6. Actual measured signal through NVISION reclocking DAs (a) Input to HD4211 reclocking DA; (b) Reclocked HD4211 output

known manufacturer's reclocking HD-SDI distribution amplifier. Output jitter exceeds SMPTE specifications.

At NVISION, we have taken great care to ensure that our PLL designs generate minimum clock jitter. This ensures that our reclocking devices provide the functionality required for the application. Figure 2-6 shows the before and after measurements, via an HD4211 reclocking HD-SDI distribution amplifier.

Our investigations have shown that the use of good reclocking DAs is currently essential, and may remain so for some time to come.

Compressed Signals

DTV signals delivered to the home are transported via a multiplexed MPEG-2 stream at about 19 Mbits (38 Mbits for CATV). This transport stream has been defined to allow the carriage of multiple video, audio and data channels. The total number of signals carried will be dependent on the compression levels and video definition and data channel requirements. The system is capable of allowing a broadcast to contain a mixture of standard definition and high definition television channels as well as new data services.

The ATSC system allows television stations to transmit received programming and insert local content via the technique of 'bit stream splicing'. The utilization of this technique requires that the local information and the original programming share identical compression algorithms and Groups Of Pictures (GOP) structures. These algorithms vary significantly with compressor design and operational settings, thus demanding that signal structure management is rigorously maintained throughout a network. Also, bit stream splicing can only occur at specific signal locations, making time accurate insertion impossible without predetermination of the desired switch point. The time required to transact a switch is determined by the compression method and the GOP structure.

For this reason many network head ends are utilizing mezzanine compression schemes to deliver programming to local stations, where the content is decompressed back to its baseband level. Some of these schemes utilize a 45 Mbit DS3 delivery service.

When considering the inclusion of an HDTV service, some operators are exploring the use of an SDI layer as a means of providing interconnect for compressed HD signals. This is certainly possible, however there are some strong operational disadvantages.

Basically, when a signal is compressed, the repetitive data within a picture sequence is discarded. In other words, if a video sequence only contained a static image, instead of repeatedly transmitting the data for each frame, it could be sent only once, together with information that describes how many times the frame is repeated. At the receiver, this frame would be received and stored so that it can be repeated locally for the duration of the sequence.

A video signal is typically dynamic; some content is static and some is in motion. Therefore, in order to compress it, GOPs are stored and compared to determine which pixels are repetitive in the given sequence, and only need to be sent once. The accuracy of compression determines the ability to differentiate between pixels that change due to signal noise, object motion, or scene changes. If a large number of pictures are stored, accuracy should increase (proided that a scene change does not occur), as more sequential images are available for comparison.

Compression research and development continues to refine our ability to accurately bit rate reduce signals from smaller GOPs. However, the complex designs required for this process are expensive. And, as it is necessary to store picture sequences, there are often considerable delays between the baseband input and compressed output. The delay time varies according to the size of the GOP. Even with I-Frame only coding, a one-frame delay is typical for the encode/decode cycle.

If this technology were utilized within a systems infrastructure, it would be necessary to surround HD recorders, effects and mixing devices with compressors/decompressors, which could be cost prohibitive. Even if the cost factors can be overcome, signal latency becomes difficult to manage. Audio paths would normally be separate from the video and with each compression and decompression cycle additional frame delay would be added to the video path. Therefore each video path would need to be paralleled by audio paths with appropriate delay compensation. (If you need them, NVISION offers digital audio delay compensators.

Several organizations have analyzed the costs and operational requirements associated with providing compressed HD via SDI paths vs. baseband HD-SDI layers and have arrived at the same conclusion. Adding the capability to distribute HD-SDI signals is by far the simplest and most cost-effective solution.

The Audio Dilemma

Designing an effective operational system for digital video signals presents some challenges, but the timing requirements and signal management techniques remain very similar to those of the analog facility. However, DTV and DVB systems bring to us a very unfamiliar audio challenge.

To date, two channels of audio in any given language have been the maximum accompaniment for any video content. DTV and DVB both are designed to deliver true CD quality, 5.1 channel surround sound to the viewer.

A large percentage of broadcasting and post facilities have been utilizing SDI video in some part, but very few have transitioned to digital audio. For that reason, we at NVISION published *THE BOOK: An Engineer's Guide to the Digital Transition* a couple of years ago. We highly recommend that you ask us for a copy if you do not currently own one. *THE BOOK* will provide you with a great deal of information on how to create and manage the AES digital audio layers that your DTV/DVB system will require. However, when *THE BOOK* was published, multi-channel audio was not a requirement, and so we will dedicate the next chapter to this subject.

Chapter 3
Multi-Channel Audio

The following piece was written by Steve Lyman of Dolby Laboratories.

Managing the Audio

Existing broadcast plants are not generally equipped to handle multichannel sound. The vast majority of plants in North America are stereo analog, which is inadequate for multichannel programming. However, all of these facilities can, and usually do, have a "surround" presence in the market through the use of Dolby Surround Pro Logic encoded programs. Dolby Surround is a matrix encoding process that allows a two-channel facility to handle four channel programs. The two-channel matrix encoded signal (often called Lt, Rt or Left total and Right total) is fully compatible with stereo analog signal paths and storage devices, as long as the channel to channel balance and phase response is consistent. Listeners equipped with stereo television receivers and a Pro Logic decoder get a Left, Center, Right and mono Surround presentation of Dolby Surround encoded programs. This allows just about any television station to have a "surround presence" in their market.

Upgrading the audio facilities from two channels of analog signal capability to an AES/EBU-based routing and distribution system makes sense. The digital signal is much more robust and is immune to many of the "finger problems" common in an analog plant. Once a signal level is set when it is digitized, it remains at that level. Machine to machine transfers are also immune to gain changes and to gain differences between the channels. Since both Left and

Right audio channels are carried in the same data stream, one half of the pair cannot get lost or have its phase inverted if it is improperly patched. This consistency makes the AES/EBU signal particularly appropriate for carrying a Dolby Surround signal around the plant.

Audio Bit Rate Reduction

One of the main differences between analog and digital television broadcasting is that the data rate of both the audio and video signals must be drastically reduced if the combined signal is going to fit the available spectrum (6 MHz or from a different point of view, 19.4 Mb/s). The audio data rate must be reduced from about 4.8 Megabits per second to 384 kilobits per second, without losing subjective quality.

The following explanation of how digital audio rate reduction works only touches on the most basic aspects of the process. It is intended to give the reader an intuitive feel for the process, rather than being an exhaustive analysis of the process. Rate reduction is all about managing quantizing noise. In linear systems, 16 bit resolution is considered to be about the practical minimum number of bits to use to keep the quantizing noise down to an acceptable level (in this case about 96 dB below the maximum signal level). If we want to use fewer bits to represent the signal, we have to find a way of dealing with the increased level of quantizing noise. Fortunately, the human hearing process provides several mechanisms to do this.

The first is the basic threshold of hearing. Our ears tend to be less sensitive at low and high frequencies than they do at mid frequencies. The second characteristic of the ear that makes rate reduction possible can be understood by considering the structure of the inner ear. The cochlea is a spiral, tapering passage with the basilar membrane stretched more or less across the diameter along its length.

Sound is conducted from the outer ear to the fluid in the cochlea where it travels the length of the basilar membrane. Different frequency components of a sound wiggle the hair cells at different locations along the membrane, stimulating the auditory nerves. The frequency dependent movement of the hair cells makes the ear act like a spectrum analyzer. A high-level frequency component will not only wiggle the hair cells at the location sensitive to that specific frequency, but some of the adjacent hair cells as well. This "spreading" of the response beyond a specific frequency can override or mask the response to other lower level, nearby frequency components. The ability of relatively loud sounds to mask lower level ones is usually described by sets of frequency and level dependent "masking curves."

If the quantizing noise produced by a coarse quantizer can be confined to the spectral region near the signal component being quantized (or encoded) and if that noise is low enough to fall below the masking curve of the signal being coded, then the listener will not be able to hear the quantizing noise.

Complex program signals are transformed into the frequency domain, and the masking curves for the different signal components computed. The masking and hearing threshold curves (and other similar phenomena) are superimposed on the spectrum of the program signal. This determines the limits on the level of quantizing noise that can be "hidden" by the program signal. The encoder can then make decisions about the coarseness of the quantizer, or the number of bits that will be assigned to each of the frequency components of the program signal.

The recovered program now no longer has the uniform low level noise floor of a PCM (linearly) coded signal, but a dynamically changing, program material dependent noise floor that is part of the program signal. The rate reduction process thus leaves its "signature" on an audio signal.

An encoder fed with a previously encoded and decoded signal will make its decisions about the amount of quantizing noise that can be concealed by that signal. The noise added by the second and subsequent rate reduction processes will add to that created by previous generations, and will rise towards the masking curve limit. At some point the demand for bits will exceed the supply, and no matter what efforts the encoder makes to avoid it (such as limiting the high frequency content) the noise will exceed the capability of the signal to mask it, and the listener will hear "coding artifacts". At this point, we can say that the process has run out of coding margin.

In very general terms, rate reduction systems that operate at low data rates do not cascade or tandem very well because they have to operate at low coding margins to achieve the low rates. Coders intended to be tandemed must operate at higher coding margins, and all other things being equal, must operate at higher data rates.

Downmixing

Television currently has to deal with one or two channel program material. Depending on the receiver, the program is either presented as such or the channels are combined for a mono presentation. Digital television is quite different in that every program will be seen by many different home receivers, each capable of presenting anywhere from one to six channels of sound, depending on the desires of the listener. The type of program and the desires of the producer will determine if the audio will be produced with either one, two, four, five or six channels. The broadcaster has no choice other than to transmit as many audio channels as are supplied by the program, with the full original dynamic range. The DTV audio system must be able to fulfill all these requirements simultaneously. This is a big change from the conventional practice of creating and transmitting a "one size fits all" program.

The key to being able to do this is to transmit some information about the audio program signal, or metadata, to the receiver. This metadata, in combination with information supplied by the listener about the number of reproduction channels available, allows the receiver to downmix a multichannel program to the number of channels available.

Control of Loudness

The current TV audio practice is to try to provide a "one size fits all" kind of signal. The mono or stereo program material is produced with a relatively restricted dynamic range that "fits into" the approximately 20 dB of headroom provided by most current systems.

Changes in loudness from program to program have always been a problem. Currently, the only way of trying to normalize the subjective loudness of programs has been to further (and sometimes drastically) reduce the dynamic range of the program material, increase the average level until all the programs occupy the top part of the dynamic range available, and are thus roughly the same loudness. This necessitates limiting or clipping the peaks to avoid over modulating the transmitter and leaves very little, if any, of the original carefully constructed program dynamics.

The ATSC sound system uses another form of metadata to provide uniform loudness to the listener. Each program style, if not each program, will have specific headroom requirements that dictate where in the available dynamic range the "average level" or loudness of the material falls. This point can be identified by the "dialnorm" metadata parameter. If the dialnorm is transmitted to the receiver along with the program, the receiver can reproduce all program material at a common loudness level. In the case of ATSC compliant receivers, the program material is attenuated by the difference between –31 dB and the dialnorm parameter. If the dialnorm value is correct, then all the program material will be reproduced 31 dB below the clipping level, and will be presented at (ideally) the same loudness.

Since there is currently no universally accepted method of measuring loudness, the process is subjective. This makes the value of dialnorm a judgement call, but also permits different styles of programs to have different loudness, as they should.

Dynamic Range Control

Listeners do, of course, need some control of the program dynamic range. Feature films, for instance, tend to have large changes in loudness which may be totally unsuited for late night listening. The best solution would be to give each listener control of the program dynamics, rather than force all listeners to make do with the same restricted dynamics as present practice does.

The choice of dynamics is also made possible with the help of metadata. The system establishes a band around the average program loudness (as defined by the dialnorm value) where no processing is done. Levels above the deadband can be reduced, and those below it can be brought up independently. This process leaves the loudness of the most important parts of the program (usually the dialog) unaffected. The listener has control of how much compression the receiver will apply so can listen to a heavily compressed program, or to the entire original dynamic range, depending on their individual desires.

The dynamic range control metadata is generated by the Dolby Digital encoder at the end of the signal chain, according to one of several compression profiles selected by the production crew. This allows selection of an artistically appropriate method of compression, rather than the one size fits all technique used today.

ATSC Audio Services

The ATSC audio system specification includes provisions for several different types of audio services. Implementation of the Associated services depends on the receiver manufacturers' willingness to supply the second audio decoder. This article describes these services for sake of completeness, but the reader should be aware that it may not be possible to provide some of the Associated services if receivers do not include the second decoder. Dual decoder receivers for special audiences may become available, but broadcasters should not count on being able to supply these additional services universally, at least in the near term. The services types are defined as follows:

Main Services

- A Complete Main service has all the elements (music, effects and dialog) of a normal complete audio program.
- The Music and Effects Main service lacks only the dialog elements. (It may also be called international sound).

Associated Services

- The Visually Impaired service can be either just a narrative description of the image, or a complete mix of all the program elements.
- The Hearing Impaired service can be supplied as dialog processed for better intelligibility or as a complete mix of all the program elements.

- The Dialog service carries one or more channels intended to be mixed into the M&E service to provide a choice of languages.
- The Commentary channel can be thought of as a dialog channel containing optional rather than necessary program contents. It may be a single channel decoded along with the Complete Main service, or may be a complete service itself.
- The Emergency service is a single channel that overrides any other service(s) that may be in use when it is transmitted.
- The Voice Over service is a single channel that is decoded and added into the center channel.

Each program item has an individual program identification code that can be selected by the listener. The Associated services are implemented by selecting either the complete or music and effects Main service and mixing the Associated Service from the output of the second decoder with the output of the main decoder. The reader is referred to section 6.6 of ATSC document A/54 for a complete description of the different service types. (See the ATSC web site at www.atsc.org).

Multichannel Layers

A straightforward transition from two channel (analog or digital) audio to multichannel audio is not an easy one. The obvious path is to upgrade existing facilities to six channels (the ".1" channel of a 5.1 channel audio path has a bandwidth of 120 Hz, but is really just another channel from the signal distribution point of view) and add a data path for the metadata. The most unimaginative thing to do in plant would be to upgrade the audio routing system to three layers of AES/EBU capability (to provide 6 audio channels) and a data layer for the metadata. The cost of this would be a lot more cabling to install, more rack space for the switchers (and jackfield) and all the operational problems of operating an additional three layers of routing in parallel.

That approach could possibly work in the plant, but a broadcast system consists of Contribution circuits that are used to bring programs or program segments together and Distribution circuits that are used to send the finished program streams to the individual stations for transmission. The majority of these Contribution and Distribution circuits are limited to something less than six channels (usually two), and do not have an associated metadata channel. The digital video tape recorders and many other storage devices form another serious bottleneck. None of the DVTRs in common use have more than four channel (two AES/EBU pairs) capability. This is a limitation of the tape formats themselves, so is not easy to overcome. Nor do they have space to store the metadata information, so cannot be used for multichannel programming.

The Contribution and Distribution circuits are a bit less limited, in that there is no mechanical media format to limit the data rates, but there are practical limitations to the bandwidth or data rate available. Satellite circuits use public spectrum and the common carriers sell "bits per second per mile" to the broadcaster. These factors suggest the use of some form of audio data rate reduction system to conserve spectrum or to reduce the cost of program distribution.

A little thought about the programming requirements makes it clear that six channels of audio is not enough. For the next several years at least, broadcasters will have to supply both the DTV and existing analog television transmitters. DTV services require from one to six channels of audio plus metadata; the analog TV service needs one or two channels of audio. The two channel "analog TV" sound tracks will probably be supplied as a Dolby Surround or

Lt, Rt signals to allow these broadcasters to have a "surround presence" in their markets, as mentioned above. Some DTV stations may also use the two-channel signal because of a (temporary) lack of multichannel facilities in their plants. Programmers will thus have to produce and distribute both soundtracks to service both markets. The Contribution and Distribution systems have to handle up to eight audio streams, plus the associated metadata

Requirements for a DTV Multichannel Audio Infrastructure

It is clear from the preceding comments that the existing signal distribution infrastructure does not meet the needs of a digital television broadcasting system. Study of these requirements has shown that the Contribution and Distribution systems have to handle up to eight channels of audio and several streams of associated metadata at a total data rate of approximately two megabits per second. There are several data rate reduction systems available, but normal broadcast operations impose several additional constraints. Program material has to be encoded for transmission, then decoded to "sweeten" or combine incoming items with locally produced items. This has to be done several[1] times as the program makes its way from the original point of production to the input to the Emission encoder. Repeated encode/decode cycles and concatenation with another type of rate reduction system (Dolby Digital in the case of DTV) usually leads to some loss of signal quality.

The most common operation in the Contribution and Distribution process is switching from one program feed to another. This may be done by switching between "live" signals, or by assembling different program segments on tape or some other storage medium. In some cases, individual items, such as an advertisement, have to be "dropped into"

[1] *A typical "worst case" number is between 6 and 9 or 10 tandems*

gaps intentionally left for them in longer program segments. This can either be done live or with insert edits done on a storage device. In many cases, the transitions between different programs are made by fading to silence, making the switch, then fading back up to the new program (a "V" fade) to eliminate any disturbing clicks or pops during the transition. Crossfades between programs or voiceovers are other types of transition, but these are generally done as part of the sweetening process or in the Master Control (Presentation) suite. Because most of the transitions made by a Master Control Switcher are tied to time of day and automated, transitions either occur or begin and end at specific time codes and hence on specific video frame boundaries.

A Rate Reduced Multichannel Audio Infrastructure

As alluded to previously, it is very difficult to replace the existing audio infrastructure with something that can handle the required number of channels and metadata. It might be possible to enlarge the audio router, but other plant equipment cannot be expanded to the necessary six or eight channels. The same idea applies to interfacility links, except in these cases, it may well be impossible to expand their capacity because of limited availability of spectrum or cost.

Some form of data rate reduction is needed to get the data rates down to practical levels for connections between and within plant facilities. Data rates must also be compatible with digital VTRs and other storage devices so that six or eight channels of audio can be recorded on these existing devices. The system selected must be able to be concatenated both with itself and with Dolby Digital without suffering any subjective quality loss. The data stream must be switchable on video frame boundaries and should produce clean (noiseless) transitions between sources. The transport stream produced by the rate reduction system should be compatible with the existing equipment found in the Contribution and Distribution chain.

Selection of a rate reduction system for Contribution and Distribution

The first candidate for this application might be the Dolby Digital system itself. It produces low data rate 5.1 channel audio, carries the metadata needed by consumer decoders, and puts all the data into an AES/EBU transport stream, so that it is compatible with all the in plant digital audio equipment. The problem with this approach is that Dolby Digital was designed as a method of delivering multichannel programs at a very low data rate. It was not intended to be concatenated, so while the degree of quality loss after several generations depends on how the program material interacts with the rate reduction algorithm, it is not possible to guarantee that some sort of coding artifact won't appear after the number of generations typically encountered in a Contribution and Distribution chain.

Dolby Digital encoders produce a complete block of rate reduced audio and metadata every 32 msec (assuming a 48 kHz audio sampling rate). This unfortunately does not match the duration of a video frame (in a television system). Any audio-follow-video transitions that are made on video frame boundaries will probably occur during a block of Dolby Digital data, and will corrupt the information. This results in a short mute (1 or 2 blocks long) at the output of the next decoder in the signal path. Thus, Dolby Digital encoders and decoders are not the ideal choice for Contribution and Distribution applications.

The Dolby E rate reduction system

Dolby has designed a new audio rate reduction system for Contribution and Distribution applications. It can be cascaded several times[2], produces clean audio-follow-video switches and carries up to eight channels of audio and the associated metadata. The design goals, outlined in the following sections, for the Dolby E system are quite different from those for the Dolby Digital (Emission) system.

Multigeneration performance

The main problem in designing rate reduction systems for multiple generations is to keep "coding artifacts" from appearing in the recovered audio after several generations. The coding artifacts are caused by a buildup of noise during successive encoding and decoding cycles, so the key to good multigeneration performance is to manage the noise optimally.

The noise is caused by the rate reduction process itself. Digitizing or quantizing a signal leads to an error signal that appears in the recovered signal as a broadband noise. The smaller the quantizer steps (ie. the more resolution or bits used) to quantize the signal, the lower the noise will be. This "quantizing noise" is related to the signal, but becomes "whiter" as the quantizer resolution rises. With resolutions less than about 5 or 6 bits and no dither, the quantizing noise is clearly related to the program material.

Bit rate reduction systems try to squeeze the data rates down to the equivalent of a few bits (or less) per sample and thus should create quantizing noise in quite prodigious quantities. The key to recovering signals that are subjectively indistinguishable from the original signals, or in which the quantizing noise is inaudible, is in allocating the available bits to the program signal components in a way that takes advantage of the ear's natural ability to mask low level signals with higher level ones.

[2] *At 1.92 Mb/s, Dolby E carrying a 5.1 channel program can be cascaded at least 50 times. Adding a stereo program or a pair of mono programs to the 5.1 channel program in the E type stream reduces the allowable number of cascades to 10. Note that these limits are highly dependent on program material, and are derived from current subjective test results.*

The masking effect of the ear can be understood by imagining the spectrum of a segment of a simple program signal consisting of a strong frequency component at, for example, 4 kHz. Add lower level signal components at 3.5 kHz and about 6 kHz. The relatively high level signal component at 4 kHz stimulates an area on the basilar membrane in the ear that is not confined to the location of the membrane that is most sensitive to 4 kHz, but tends to spread along the membrane. The two lower level signals will also stimulate the basilar membrane, but their stimulus may be overcome, or masked, by the higher level signal, thus rendering the them inaudible[3]. Now if the quantizing noise associated with each program signal component can be confined to the region of the spectrum that is masked by that component, and if the noise level is not allowed to rise above the "masking threshold" it will be present, but inaudible in the recovered signal.

The rate reduction encoder sends information about the frequency spectrum of the program signal to the decoder. The set of reconstruction filters in the decoder confines the quantizing noise produced by the bit allocation process in the encoder to the bandwidth of those filters. This allows the system designer to keep the noise (ideally) below the masking thresholds produced by the program signal. The whole process of allocating different numbers of bits to different program signal components (or of quantizing them at different resolutions) creates a noise floor that is related to the program signal and to the rate reduction algorithm used. The key to doing this is to have an accurate model of the masking characteristics of the ear, and in allocating the available bits to each signal component so that the masking threshold is not exceeded.

[3] *Note that there is no one single "ear model" that is universally accepted, and that an algorithm designer may have refined the various published models, based on their experience in designing and testing rate reduction systems.*

When a program is decoded then re-encoded, the re-encoding process (and any subsequent ones) adds its noise to the noise already present. Eventually the noise present in some part of the spectrum will build up to the point where it becomes audible, or exceeds the allowable "coding margin". A codec designed for minimum data rate has to use lower coding margins (or more aggressive bit allocation strategies) than one intended to produce high quality signals after many generations

The design strategy for a multigeneration rate reduction system, such as one used for Dolby E, is therefore quite different than that of a minimum data rate codec intended for program Emission applications.

Switching

The most common operation in the Contribution and Distribution of television signals is a simple cut transition between program segments. Other operations, like V fades (fade to silence then back to unity) are placed around a cut transition to ensure that here are no transients at the transition. The majority of audio cut transitions are made at the vertical interval switch point of the video signal because it is convenient to slave the audio switcher to the video switcher and because the transition point can easily be labeled with video time code. Unfortunately, as was pointed out in the case of Dolby Digital coding, the block structure of rate reduced audio systems does not match the video frame structure. Audio follow video switches almost inevitably corrupt the blocks of data and cause some sort of interruption in the recovered audio.

The situation for baseband (or PCM) audio is not much better. The majority of digital audio switchers are simple "crash" switchers that make the cut as soon as they receive a command, so can corrupt the audio sample structure. Synchronous switchers wait until the beginning of the next audio sample pair (AES frame) before making the cut, so do preserve the data integrity, eliminating one source of transients. Unfortunately, audio transitions may fall at a time when a large peak of one polarity in the first signal matches a peak of the opposite polarity in the other signal. This produces a sharp transient click in the resulting signal, so even a synchronous switcher cannot guarantee clean transitions between audio signals. V fades that make the transition during the silent period are the usual cure for this problem.

The Dolby E system design assumes that audio transitions will take place at vertical interval switching points[4], so aligns the blocks of rate reduced audio data with these points. The rate reduction algorithm sacrifices a small amount of coding efficiency so that the decoder can make short cross fades between the end of one block and the beginning of the next block of audio information. This eliminates transients at switching points, even if the "positive peak to negative peak" problem is present at the transition point and produces reliably clean transitions. The switching capabilities of the Dolby E system were demonstrated during the 1998 NAB Convention. Attendees were able to switch freely between program streams from three different digital VTRs while listening to the result on a high quality multichannel monitoring system.

Metadata

As mentioned in the introduction, metadata is an essential ingredient of the ATSC audio system. The unfortunate part of the existing signal distribution system in existing plants is that there is no signal path for the metadata. The Dolby E system carries metadata in a multiplex with the rate re-

[4] *The switch point and video frame duration are different for different video systems, so the E type encoder uses the video reference signal ("color black" or its equivalent) of the video system associated with the audio to generate the clock and timing information it needs.*

duced audio, thus providing a way of moving metadata through the Contribution and Distribution links.

Dolby E also caries up to eight channels of audio. Any one of the eight channels, or any combination of up to eight channels can be defined as a program, and thus have a group of metadata parameters associated with it. The most common combination for DTV applications will probably be a group of six channels (for the main 5.1 channel program) and a left, right pair carrying a Dolby Surround (Lt, Rt) signal for the associated analog TV service. In this case, there would be two groups of metadata in the multiplex, or one for each program service.

Most of the metadata carried by the Dolby Digital Emission system is intended to allow individual listeners to tailor the audio presentation to their needs, so is referred to as consumer metadata. The Dolby E system also carries Professional Metadata that can be used by the broadcaster to resynchronize, monitor and modify the level of the decoded audio, again on a program by program basis. The professional metadata is only used in broadcast operations, and is never sent to the home DTV receiver

Time stamping

Time stamps are an important part of the Dolby E data stream. Many operations require treating the audio and video portions of the program individually; it would be a pity if there were no convenient way of reliably laying the audio back, in sync with the video. SMPTE Time Code is fed to the E type encoder and multiplexed into the data stream so that it can be recovered by the decoder. It is intended to be a time stamp, rather than a way of keeping track of time of day, so the recovered time code is identical to the time code that occurred when the audio was being encoded, and does not take into account any encoding or decoding delays. The Drop Frame flag and user bits are also carried.

Monitoring program signal levels

A common operation, particularly in the Distribution part of the signal chain, is to monitor the level of the program signal. In areas where many programs are present simultaneously, it might be too confusing to reproduce each program from its own set of loudspeakers, or not timely enough to switch one set of speakers between the various programs. Level monitoring can provide some level of confidence that the program material is still present, and can be done for many feeds at the same time without confusing the operator.

It would be a pity to have to decode the Dolby E data stream just to drive a set of meters to indicate that there was some activity in each of the channels of a program group. Part of the professional metadata is metering information. The individual channel signal levels are measured during the encoding process and carried in the professional metadata. Measurements are of the peak and RMS signal levels over the entire block duration (or during one frame period of the associated video reference signal). The amplitude resolution of the measurements is approximately 0.1 dB. There is no attempt to provide a combined signal level for each program group.

Changing levels

The beauty of digital audio is that levels stay the same. Unlike the analog days when the common carriers guaranteed signal presence but not its level, digital data is not expected to change between the transmission and reception points; the green tweaker that hung on the equipment racks can be retired. But there are still some good reasons for being able to trim a signal level, so the Dolby E professional metadata carries gain words that can instruct a decoder to change the level of a received signal. Each block of data carries two gain words, one applicable to the beginning of the block, and one that applies to the end of the block. If they are different, the decoder interpolates a linear ramp over the duration of the block (or over 1 frame of the associated video) to avoid "zipper noise" as the level changes. The gain range is from +6 dB to minus infinity and is applied to all channels in the program group equally.

The ability to change the level of the recovered program signal is probably more useful as a way of doing fades without loosing a generation than it is as a gain adjustment. As mentioned earlier, the V fade is a very common transition between programs and because it is used on air so often, is usually initiated by a presentation automation system that specifies times and event durations in units of time code. Metadata gain words have the same temporal resolution, so the whole concept fits very easily into normal operational practices. Gain words within a few frames of the end of one program segment instruct the decoder to ramp the level of the program down to silence, the switcher cuts to the next program segment whose gain words cause the decoder to ramp the level back to unity during another few frames.

The Transport Mechanism

The key to making the E type rate reduction concept practical is to make it easy to integrate with the existing broadcast plant infrastructure. It is a relatively high rate digital signal, so will not integrate easily with an analog plant. As discussed in the Existing Infrastructure section however, DTV requires at least digital VTRs, all of which are currently limited to recording one or two AES/EBU digital audio signal pairs. If at least one layer of AES/EBU signal distribution and routing capability can be added to an existing plant (which may well already have this capability) and if the Dolby E signal can be transported by the AES/EBU mechanism, then that plant immediately becomes capable of doing multichannel programming.

The AES/EBU signal carries two audio sample words, each in its own audio subframe, during each sample period. The two subframes start with a 4 bit Preamble or sync word which is followed by 24 bits of audio data payload space. The subframes end with four additional bits (one for each of the Validity, Channel Status, User and Parity bits) for a total of 64 bits for both subframes. Bit 1 of Byte 0 of the Channel Status information can be set to indicate that the information carried in the audio payload space is not an audio signal. We are thus free to put as much non audio information as we care to in the payload space.

The most sensible choice for the time being would be to use the 20 MSBs[5] of this space for Dolby E data, as this is the maximum number of bits that can be recorded on most studio level digital VTRs. Fortunately this produces a data rate of (20 + 20) x 48 kHz = 1.92 Mb/s which is quite sufficient for Dolby E data. Note that the number of bits used to transport the Dolby E signal has nothing to do with the dynamic range of the audio signal carried by the Dolby E system. The current specification of audio program dynamic range is 110 dB, or the equivalent of 18 bits.

The other advantage of using the AES/EBU signal as a transport mechanism is that the Dolby E signal immediately becomes compatible with the rest of the digital audio equipment in the plant. It can be switched, recorded, edited (cuts or insert and assemble edits) just like any other digital audio signal, as long as some basic precautions are observed. The data must not be changed by any part of the system it passes through as this would destroy the coded audio information (and the metadata). Specifically:

1) Any gain controls must have a unity gain position or bypass function that ensures that the data recorded

[5] *If the entire Contribution / Distribution system was transparent to 24 bits, the increased data rate could be used to improve the mutigeneration performance of Dolby E. 16 bit wide data paths are also acceptable, but may limit the number of channels available, or the number of artifact free generations that can be expected.*

by or passing through the system is an exact duplicate of the input data.

2) A system must not change the word length of a non-audio signal by truncating it. If the data is being carried in as AES/EBU data stream, the channel status information (byte 2, bits 0 to 5) should be set to indicate the intended word length. The channel status information should either be carried through the system or set appropriately at the system output.

3) Recording systems, switchers, editing systems and similar devices must be able to make butt splices in data streams. The switching points must happen during the vertical interval switching period of the video used as the sync reference to avoid destroying the encoded audio data.

4) Any cross fades, fades to or from silence, sample rate conversions or other process that are intended to modify the data (including rounding or dithering) must be bypassed when handling non-linearly coded data

5) Dolby E rate reduction systems assume an error free channel between the encoder and decoder(s). Any channel coding intended to protect the E type data from error prone channels must be provided by the communications channel in use.

Application of Dolby E type codecs to Program Contribution and Distribution

The previous sections have covered the basic requirements of a rate reduction system intended for Contribution and Distribution applications. Figure 3-1 shows a few of the areas where the Dolby E codec can be used.

Mobile or Outside Broadcast vans are often used where no leased lines are available and have to rely on a microwave link. Digital radios provide data rates from about 20 to 50 Mb/s, depending on what type of modulation is used. There are claims that "there will be a need for a variety of digital modems and multiplexers" to marry the different video and audio sources to the microwave system in use. Total data rates like these will allow space for approximately 2 Mb/s of Dolby E encoded audio without squeezing the video channel seriously. In some cases, digital radio systems also provide a "wayside T1" connection that provides a 1.544 Mb/s connection that could be used to carry the E type data at 1.536 Mb/s, which is the data rate of the E type information when transported in the 16 MSBs of the AES/EBU stream. This is also the version of the data that will fit on the 16 bit digital audio tracks of some ENG cameras, turning them into multichannel capable devices.

Remote studio operations are generally more entrenched, so would tend to use leased facilities to haul the program signals back to a studio center or a network origination point. In this case, the audio and video will probably share a 45 Mb/s DS3 service. The baseband audio data rates are in the order of 5.5 Mb/s which would take too much of the channel away from the video signal, but at about 2 Mb/s, the Dolby E data can comfortable share the channel with the video signal.

Satellite distribution facilities can also provide data rates from about 45 Mb/s to 60 Mb/s, depending on the transponder bandwidth. The same ideas apply here as for the remote studio situation.

Post production facilities have to find a new release format for finished high definition, multichannel programs. The emergence of mezzanine level video rate reduction codecs allows existing tape formats to carry HD material, but the tape format is unchanged and limited to four audio channels (or two AES/EBU pairs). E type encoders could allow these tapes to carry up to 16 channels of audio and metadata, but in practice, probably only one track pair

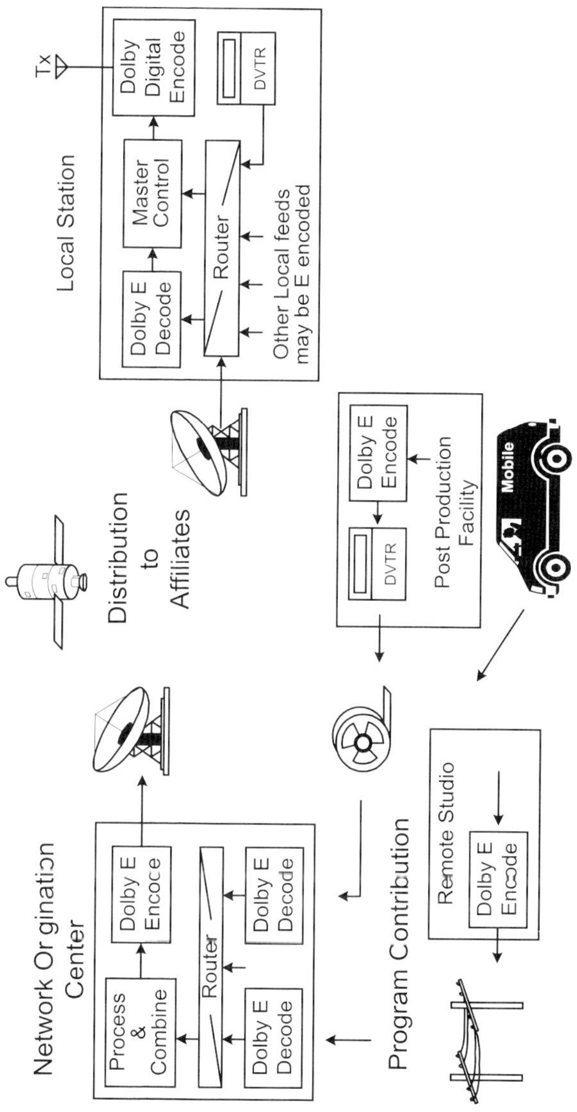

Figure 3-1. Application of Dolby E codecs to Program Contribution and Distribution

will be used for E type data. It will carry a 5.1 channel program (with its metadata) and an Lt, Rt Dolby Surround version of the program, with metadata, on the remaining two channels. The second AES/EBU track pair will be used to carry another Lt, Rt version of the program that has been mixed for NTSC release. This is particularly convenient for stations that have no DTV service or digital infrastructure, as they will be able to take the analog output of the "NTSC" track pair and operate as usual. This "multiservice" release format is particularly well suited to advertisements and programs with high production values, as it allows to program producer to tailor the audio to the intended service, but uses only one common tape format to carry them all.

Dolby E thus provides a "point of production to transmitter" path for multichannel audio and the necessary metadata, using the same codecs operating at the same data rates serve all the applications.

The preceding piece has been reprinted with the permission of Dolby Laboratories.

An alternative to compressed audio layers

While Digital Dolby and Dolby E systems obviously provide excellent methods of distributing and storing multichannel audio via currently available equipment, any form of compression requires additional processing equipment that of necessity carries a cost and operational overhead.

At NVISION we believe that where an audio origination facility is to be included within the plant infrastructure, a method of accurately managing baseband multichannel audio is required.

To achieve this, some vital limitations need to be addressed. As previously mentioned, a major limitation to managing baseband multichannel audio associated with a video source, is the limited number of channels available on currently available recorders. Most DVTRs have a maximum of 4 audio channels, therefore unless separate multichannel audio recorders are slaved to each video machine, distributing baseband surround sound is currently impossible. Obviously, managing slaved audio recorders doubles the complexity of machine and resource management, although some organizations have elected to adopt this methodology.

If DVTRs were capable of more channels, say eight, then the next problem would be signal distribution. Four layers of AES audio are perhaps more easily managed than multiple DVTRs and DA88s, but the possibility of phase slips between AES streams and varying path lengths could also present a difficult challenge.

Accordingly, NVISION proposed a new audio transport method to the SMPTE organization, following discussion with various other equipment designers and manufacturers. At the time of this writing, this proposal has been approved by the SMPTE standards committee, but has not yet been ratified or assigned a standard number. It is anticipated that this standard will be published by mid 1999. The following is a synopsis of this standards proposal.

A Proposal for Transporting Multi-Channel Audio

The advent of ATV, be it HDTV, resolution enhanced SDTV, multi-cast, or multi-lingual, has generated increased desire to improve audio quality. In fact, results from a number of ATV user tests indicate that improving audio quality provides at least as much enhancement to the viewing experience as increasing picture aspect ratio and resolution. The surround sound or theatre sound environment can be implemented and coded in many forms. AC-3, DTS,

SDDS, MPEG Audio, and Pro-Logic are some examples. Each of these multi-channel coding methods requires original audio source material that is image accurate both during production and for production master distribution in order to maintain the original spatial perception of the program.

It is presently difficult to maintain proper audio image in a video production environment. One need only investigate the past years of effort on the part of AES and SMPTE technical committees and the lack of even a recommended practice, to realize the challenging nature of this problem. Recently, several proposals have been made for distribution of audio using compression. While compression is very useful for recording, long haul transmission, and final delivery, it creates some problems as a distribution method for production. Production requires that multiple sources be available to a common processing point, and that these sources be in a common editable format. Since there are a number of compression formats available, the point of production must be "multi-lingual", and must also be able to manage the different time delays associated with each format, and the variable time delay with a given format. This is an expensive process to have distributed throughout a production facility, and many system integrators who have looked seriously at this implementation have realized that it is expensive and impractical for both audio and video.

DVTRs are not transparent devices in spite of the fact that they are essentially data recorders. There are several audio signal impairment issues including phase uncertainty between AES3 audio and video sync. Error concealment algorithms, sample rate converters, and level processing are operations that may be found in many machines. If compressed data is stored on the machine, the effects of these processing steps, if not successfully bypassed, will typically render the uncompressed audio output useless. Different machines may well require different coding algorithms to insure transparency within the recorder. As a minimum, a common, or standard, compression algorithm would need to allow for full error correction given the statistics of every transport, fixed compression time, across all equipment in the plant, and a coding scheme which would allow a cut edit. It is most likely that future machines will employ proprietary compression to accomplish the feature sets desired by individual machine manufacturers. Furthermore, any saleable DVTR (or disk recorder) will necessarily have full bandwidth outputs for both video and audio.

NVISION has proposed to SMPTE, a straightforward intra-plant distribution solution based on multiplexed 12-channel full-bandwidth distribution and switching between equipment with CODECs (encoder-decoder pairs), such recorders or STL interface points, where compression may be required for bandwidth preservation. This topology insures that digital audio signals are coherent and that image accurate audio may be moved throughout the facility, with confidence. It also allows for optimal CODEC design to match the bandwidth and error behavior of specific channels.

Six AES3 signals may be easily transported in a single path for only a small increase in cost over a single AES3 signal. The cost of the multiplexer and demultiplexer is minimal when compared with separate distribution CODEC costs. Given the proliferation of 8 track audio recorders, the presence of 8, 10, and 12-channel HDTV DVTRs, and the economies associated with multi-channel transport, a 12-channel interface standard will dramatically lower the cost of in-plant audio distribution.

The Multiplexed Transmission Format

The multiplexed serial data format calls for transmitting consecutive data frames composed of twelve AES3 data packets and one Header Packet, in the rigid order shown in figure 3-2. Note that the AES3 inputs or outputs of a multiplexer or demultiplexer have a rigid position in the transmission frame format. This is a critical economic and operational advantage for manufacturers and users.

Each AES3 packet is a truncated version of the 32 bit AES3 sub-frame as shown in figure 3-3. This packet is 28 bits long, with complete preservation of the entire AES3 sub-frame data payload. The block start information is moved to the multiplex header for data efficiency.

0 47	48 75	76 103	104 131	132 159	160 187	188 215	216 243	244 271	272 299	300 327	328 355	356 383
Header	Ch 1	Ch 2	Ch 3	Ch 4	Ch 5	Ch 6	Ch 7	Ch 8	Ch 9	Ch 10	Ch 11	Ch 12

Figure 3-2. Channel packet structure

The header packet contains 48 bits. See figure 3-4. A four-bit preamble is used for multiplexer framing. A single MC bit is used for alignment of multiple data streams insuring synchronous performance and preservation of audio image. The header contains 4 bytes of channel data, an op-

24 Bit Audio or 20 Bit Audio + 4 Bit Aux	V	U	C	P

Figure 3-3. Twelve-channel sub-frame

tional channel block bit, and two reserved bits. A parity bit (CP) sets this group to even parity. The last byte of the header contains the MC bit, the Z bits corresponding to each of the AES3 inputs, and a second parity bit which sets the last byte to even parity. The Z preamble from each pair of AES subframes is saved as a Z bit for decoding channel status. The Z bit will be 1 at the start of the AES3 block, and 0 for the remaining 191 sub-frames in the standard AES3 block. This allows accurate recovery of all the channel status information. The Z bit is coincident in time for the AES3 A/B pair since this is a requirement of the AES3 specification. Equipment is required to pass channel status transparently. Also, equipment that processes audio and reinitializes the channel status bits must restripe the Z framing bit in accordance with maintaining the channel pair correlation.

The complete multiplexed frame of data is 384 bits long generating a bit rate of 384 x FS, the audio sampling frequency. For example, 48 kHz audio is transmitted at 18.432 MHz.

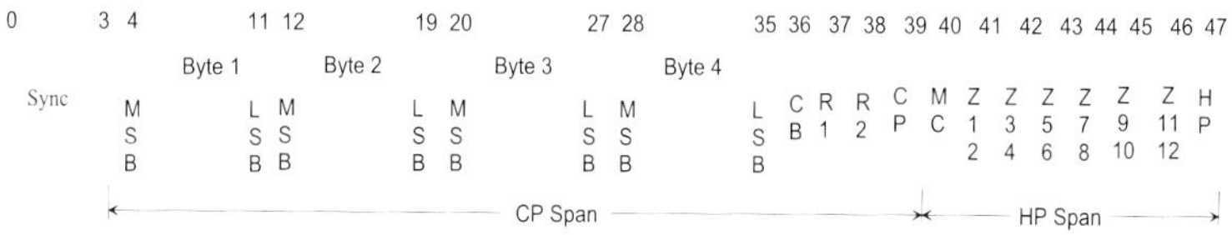

Figure 3-4. Twelve-channel header packet

Channel Code

The multiplexed data stream utilizes Bi-Phase mark coding. Sync is provided by a code violation consisting of four baud periods low followed by four baud periods high (or the inverse). The advantage of this coding scheme is that the clock is always contained in the data, and easily recoverable. Also, the code violation simplifies framing location, the channel code is insensitive to polarity, and the required circuitry is inexpensive.

Electrical Interface

The electrical interface is coaxial with a 75 Ohm characteristic impedance. Connections are made with BNC type connectors, and signal lines are terminated with 75 Ohms. The output signal level is 1.0 V peak to peak ±10%. Rise time is 5% to 30% of the baud interval. A 3 ns rise time will stay within specification for sample rates from 32 kHz to 96 kHz. This signal is very robust. Since the energy band is largely above the RC turnover frequency in typical coaxial cable, the need for cable equalization is reduced or eliminated in most situations, and jitter introduction is minimal without equalizers.

Postscript

The utilization of this standard will depend on the manufacturers of storage and editing equipment, providing equipment with a sufficient number audio channels and the appropriate multiplexed interface. This is currently under consideration by many, and if adopted will provide an inexpensive and manageable alternative to compression, without any of the associated editing and signal latency difficulties.

Chapter 4
Routing Systems

The routing system is the heart of any television plant. It provides the interconnect for all of the video, audio, time code and data signals that are required for normal operations. Over time, the normal operational requirements have significantly increased the demand for larger and more complex switching matrices. A few years ago, a 64 x 64 video router was considered to be very large; today it is considered as average. As DTV facilities are constructed, the demand for larger and more flexible systems grows.

There are two major components required to create and operate an effective signal management system: the control system that provides the switching intelligence, and the signal routing matrices. The choice of these two elements is critical in designing an effective and flexible means of signal management that remains fully functional for many years.

Most of us know that making a major change to an operational plant's routing system, without interrupting normal operations, can be very difficult, if not impossible, In comparison, it is relatively easy to add new editing equipment, or to replace tired tape machines with ones that offer the latest technology. This statement is born out by the fact that there are routing systems still in service that were installed many years ago.

In the analog domain, NTSC and PAL signals have remained essentially unchanged since their introduction. Of course the technical quality of these signals has improved over time, but an effective routing system built twenty years ago can still operate successfully (if it can be maintained).

Any system built today should be designed to remain in operation for many years. This is a much bigger challenge than it may initially seem. Earlier in this book, we mentioned the variety of possible picture formats and signal types that may need to be included within a DTV system. Today's system needs to provide an easy path to add signal types that may not be required initially. Further, with the potential need for multiple signal types, control systems need to provide a far greater level of intelligence and flexibility than before.

The next chapter is dedicated to control system topology and deployment; in this section, we will discuss the switching matrices.

Switching AES/EBU Digital Audio

Matrices designed for digital audio fall into two basic categories, asynchronous and synchronous. Asynchronous switches are the most commonly available from traditional router manufacturers. (Unless otherwise stated, AES audio routing is asynchronous). Synchronous switches are analogous to video vertical interval switching, in that they align AES/EBU audio frames to provide frame accurate switches.

Asynchronous switching is ideal for any environment where mixed audio sample rates are necessary and the router is used only to preselect sources, rather than for live or 'hot' switching. An asynchronous switch is basically a 'crash switch', in that it will switch signals regardless of timing relationship. If used live, this type of switch will cause disruptions in the output audio framing that will generate audible switching artifacts. These errors will often cause the downstream receiver to lose lock and the audible effect will vary according to receiver design. This effect can vary from output muting until lock is reacquired to very objectionable crackles, clicks and pops.

For on-air or live switching, synchronous switching is necessary. Synchronous switching is achieved by ensuring that all audio sample rates (e.g. 48kHz) are the same and are locked to a common reference. The routing switch frame must align all audio samples at the input to the matrix so that a switch is performed at a frame boundary. See figure 4-1. New audio data is then inserted into a continuous AES/EBU audio stream at the output stage. This technique eliminates receiver errors and was pioneered by NVISION. NVISION synchronous audio routers are now the most widely used products for on-air applications and are available in sizes from 8 x 32 to 2048 x 2048.

Digital audio signal referencing and management is neither complicated nor expensive, but it is easy to make mistakes within a system design if the nuances are not well understood. As such mistakes can be difficult and expensive to overcome after installation, we recommend that the NVISION engineering guide, *THE BOOK*, be used as a reference when defining audio layer designs.

Routing Multichannel Audio

All NVISION asynchronous and synchronous audio routers will pass Dolby Digital (AC3) and Dolby E compressed audio signals. These signals must be switched synchronously and coincident with vertical interval to avoid receiver error. However, your environment may require that multichannel audio is managed at baseband levels. For this application, NVISION synchronous routing provides a very sensible cost effective solution. Our large, expandable AES routers provide the ability to include four audio

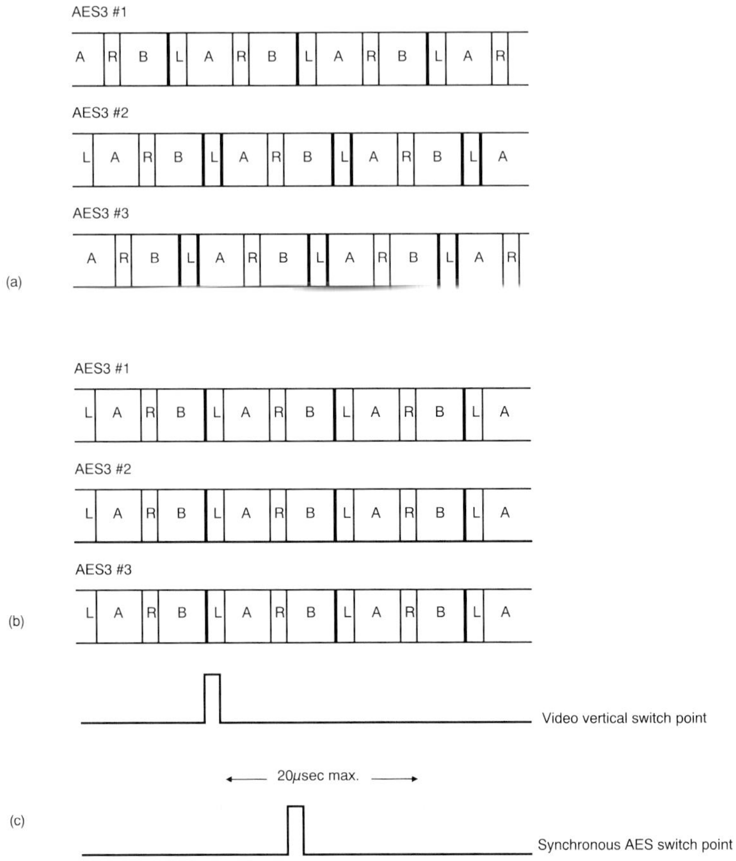

Figure 4-1. AES input reframing for synchronous routing

layers within the same switch. These layers may be switched separately or linked together to ensure that four AES streams (eight channels) can be switched simultaneously. See figure 4-2.

Data Routing

Data routing has been employed primarily for machine control applications. In recent years, data routing has become common in larger facilities. However, as serial com-

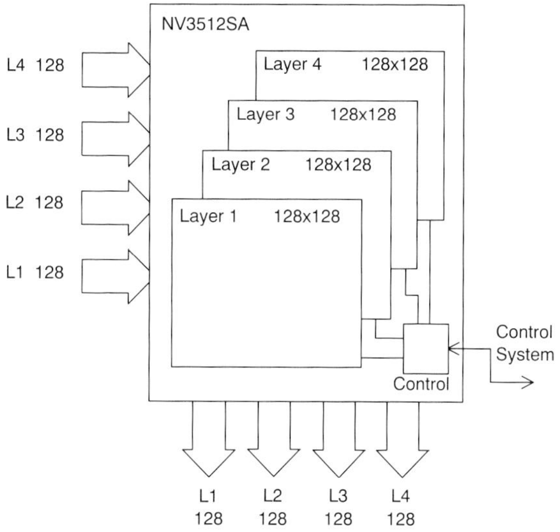

Figure 4-2. Four discrete audio layers within a single NV3512

munications (RS232, RS422 and RS485) are bi-directional in nature, routing is not as straightforward as with video or audio. Ideally, router designs should allow data ports to be dynamically configurable to the task required, i.e. a VTR can be controlled or its control panel can be used to control another device. In this circumstance the VTR's serial port send and receive lines swap, when the machine is switched from remote to local. The data router should be capable of mirroring this change. Otherwise, two router ports with opposite pin connections would be required for this level of flexibility.

In instances where dual router ports are used to allow a VTR to be configured as controlled or controlling, conflicts can occur within the router that will prevent communication completely. NVISION manufactures data routing systems that utilize a patented 'Dynamic Port' (fig. 4-3) design that overcomes these problems and provides the operator with total flexibility. (This technology is fully described in *THE BOOK*).

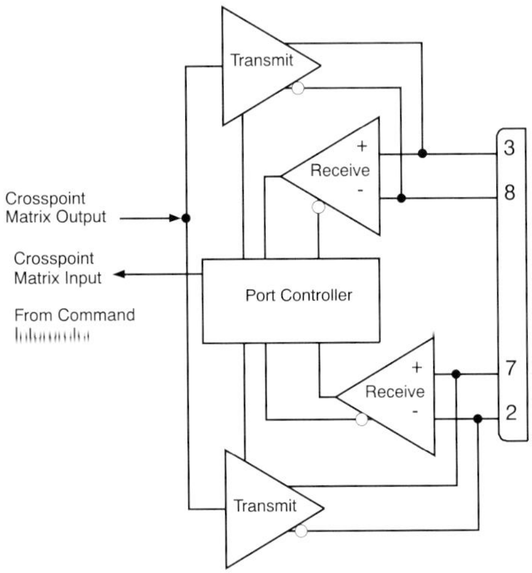

Figure 4-3. NVISION's patented dynamic port architecture

The metadata and data services that may be required in your DTV plant can also be routed via this type of switch. This makes it possible to link specific audio signals to their associated metadata, where necessary.

Time Code Routing

Historically, time code has been treated as an analog signal for the purpose of distribution and routing. However, time code is a digital signal that has a relatively low bandwidth and so has been traditionally routed via analog audio switches. Using analog technology to meet this requirement is not cost effective today. Digital technology makes the design and manufacture of a digital time code router simple and inexpensive. NVISION manufactures several all-digital time code routing switches (from 8 x 32 to 512 x 512) that are generally more reliable and less expensive than other manufacturers, analog offerings. They provide the additional benefit of reshaping signal edges, allowing time code to be accurately read during fast and slow VTR shuttle speeds. (For further information, please refer to *THE BOOK*)

Video Routing

For any of the routing layers (video, audio, timecode, data), allowance for future expansion is paramount in ensuring a system's longevity. Systems with limited expandability have been a consistent cause of frustration, often requiring the use of tandem systems or overworked patch bays to overcome size limitations, before total replacement becomes practical.

For the new DTV video formats, a decision to implement a routing switch that is fixed to signal type could create an operationally inadequate plant of the future. We have spent a great deal of time considering and researching this issue, with the following results:

1. Router Size and Expandability

Most SDI router designs developed as an extension of analog routing architecture. Switch sizes beyond 128 x 128 were once considered unmanageable and so most router designs are not linearly expandable beyond this. Expansion for larger switching requirements often uses 'tie lines' between switching blocks. Tie lines need very careful planning and management if bottlenecks are to be avoided.

Users requiring larger matrices usually need to expand these routers geometrically. To create a matrix that is 256 x 256, using 128 x 128 switches, with all inputs available to all outputs, requires that a minimum of four 128 x 128 (128^2) frames be interconnected and all 256 inputs be distributed across the four frames. See figure 4-4. This method is expensive and requires the provision of a large amount

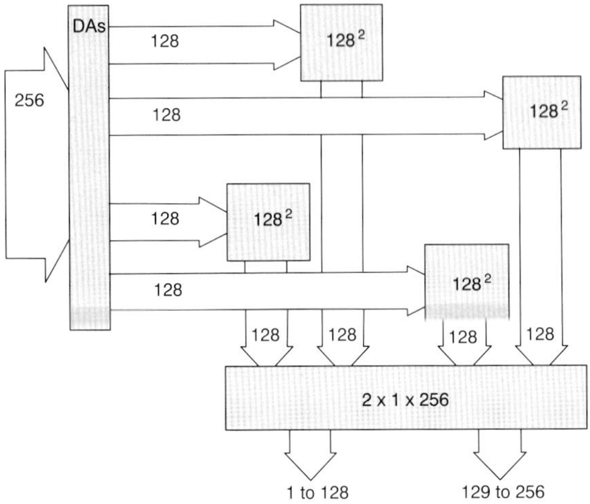

Figure 4-4. Traditional geometric router expansion

of rack space. The larger the base matrix size, though, the more acceptable the cost and rack space usage becomes.

2. Signal management

Router designs are usually dedicated to managing a specific signal type. Most SDI routers will pass all of the SDI data rates, but may not provide reclocking (which is essential in a large facility). Generally they are capable of providing vertical interval switching at only one frame rate (although some can manage dual frame rate vertical interval switching (59.94 and 50Hz), provided that the switch is partitioned appropriately).

As the SDI signals are NRZI coded, the use of all of the outputs of differential amplifiers within the design is normal. Therefore, output signals may or may not invert, depending on the internal signal path (which may be an issue for passing ASI signals—see chapter 2)..

3. High Definition Signals

Traditional mother- and daughter-board architecture, necessary to provide ease of service, precludes the possibility of adding serial high definition signals. At 1.5 Gbits, signal path impedance and length must be carefully controlled. Even the connectors normally employed will provide unacceptable impedance differences, making it impossible to successfully pass these frequencies.

Accordingly, manufacturers have developed alternative architectures to manage these signals. Many of these designs restrict the matrix size to eliminate the need for mother and daughter boards, providing direct connection from the I/O to the crosspoint. These devices are normally limited to 16 or 32 square.

Other designs for larger sizes make the input and output cards accessible only from the rear of the frame, making online servicing almost impossible.

Engineers can readily estimate their current SDI switching requirements, and can probably make a fair estimation of future ones. For HD-SDI, a need for small switches currently exists and future requirements are difficult to predict.

Our Conclusions

A new type of routing switch architecture was required:

1. Size requirements are continuing to expand as new services are added. A more readily expandable switching system is necessary.
2. The need to switch in vertical interval at several frame rates may be a requirement of any DTV system. No paths should invert, so that ASI signals (NRZ coded) can be reliably managed.

3. A design that can competently manage HD-SDI signals in a modular, expandable and serviceable manner is required.
4. Ideally, all digital video signals should be managed within the same switch design.

Since drawing these conclusions, NVISION has designed a new line of digital routing switches, called ENVOY. These products accommodate all of the requirements that our market research has identified, plus a few others.

Universal Digital Video Routing

In order to meet the present and future needs identified for the DTV plant, a new routing switch architecture was required. That architecture needed to provide the ability to sensibly manage a multiplicity of data rates, vertical intervals and matrix sizes, affordably.

We decided to design a product that would manage HD-SDI data rates and could also be used to switch lower speed signals. The product must provide this functionality without a significant increase in purchase cost over traditional SDI switches.

The heart of any switch is the crosspoint matrix and an expandable switch requires the ability to add multiple crosspoints. Designing a fixed size crosspoint to switch data rates at HD-SDI frequencies is challenging but requires care rather than invention. At NVISION, we feel strongly that any routing system design must offer operational longevity and flexibility. Accordingly, our engineers were commissioned to design a crosspoint matrix that could: a) pass data-rates from DC to in excess of 1.5 Gbits, b) allow the addition of multiple crosspoint modules for easy expansion, and c) provide the ability to switch on different vertical intervals at the same time.

A new backplane or motherboard construction was needed to provide the interconnectivity between modules at very high data rates, and at the same time provide the ability to 'hot' swap modules for service.

A selection of input and output modules are necessary to allow cost effective inclusion of the various signal types. HD-SDI inputs and outputs require the use of more expensive components than for SDI signals and so modules for each signal type were developed so that an SDI switching systems would be competitively priced, without the burden of HD-SDI circuits. If HD-SDI signals are required the additional costs should still remain competitive with other HD switches.

New control intelligence was also required to allow the mix and match of vertical intervals, as well as to allow these routing switches to be operated under other manufacturers, control systems.

The culmination of this research and development is the ENVOY series of universal digital video switches. The first models available are:

ENVOY 6064 (fig. 4-5)

This 64 x 64 frame has a 64 x 64 crosspoint card that is fully redundant with the addition of a second card. This second card is a hot standby and fully mirrors the 'live' card. In the event of a failure, the standby can be instantly accessed to ensure that operations remain uninterrupted. This changeover can be carried out manually or automatically (dependent on control system). Input cards are available for SDI and HD-SDI formats and each card adds eight inputs (eight cards, for a total of 64 inputs). Naturally, both types of inputs can be included at the same time. Output cards also provide for SDI or HD-SDI and have eight dual outputs per card (16 connections). A redundant power supply can also be included.

This switch is ideal for small to medium applications where flexibility is required, without the need of expansion beyond 64 x 64.

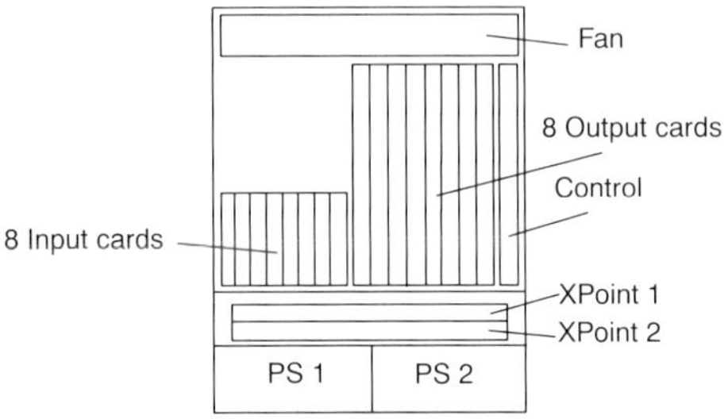

Figure 4-5. ENVOY 6064 module layout

ENVOY 6128 (fig. 4-6)

This 128 x 128 frame shares the same cards and features as the ENVOY 6064, except that it utilizes a different crosspoint card design. The crosspoint card is sized 128 x 32; four cards are required for a fully loaded switch. There is no hot standby function, but the output impact block is limited to 32 outputs.

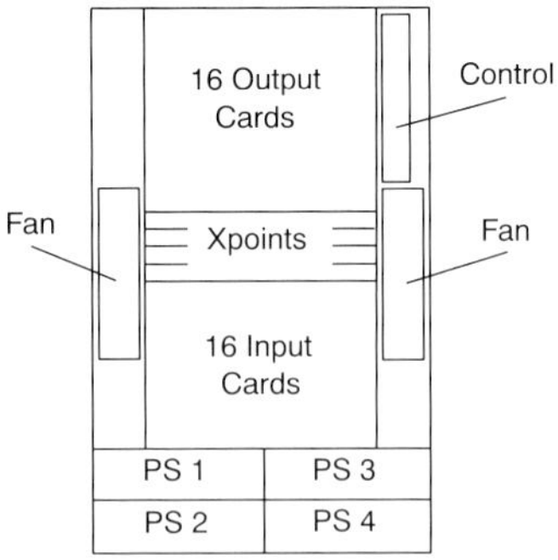

Figure 4-6. ENVOY 6128 module layout

ENVOY 6256 (fig. 4-7)

This 256 x 128 frame utilizes the same cards as the ENVOY 6128 and is designed for large and expandable systems. This switch can easily and cost effectively be expanded to 256 x 256 with the addition of another frame and NVISION DAs. See figure 4-8.

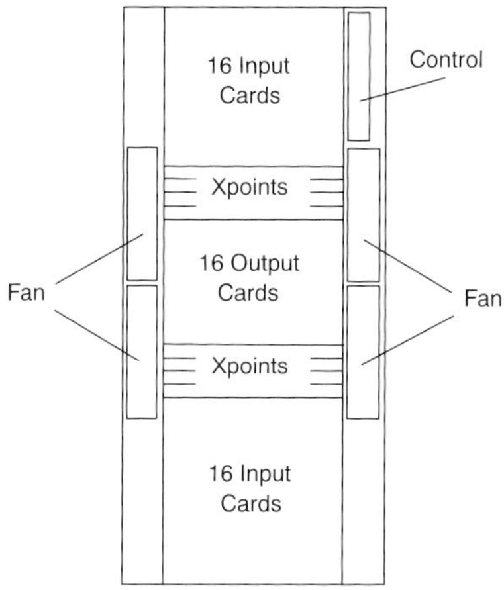

Figure 4-7. ENVOY 6256 module layout

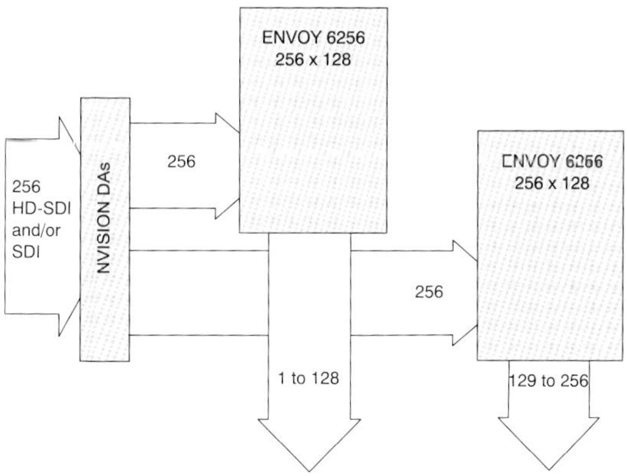

Figure 4-8. ENVOY 6256 expanded to 256 x 256

ENVOY Features

- All primary outputs are non-inverting, for the inclusion of ASI (NRZ coded) signals. The secondary outputs are inverted.
- All outputs are reclocked, ensuring that technical signal quality is maintained.
- Switches SDI and HD-SDI signals, in the same frame, at the same time.
- Supports switching on any standard vertical interval (60/59.94/25/24/23.976. I or P).
- Time code inputs (LTC and VITC) for determined switching, dependent on control system.
- Comprehensive on board diagnostics for health monitoring, dependent on control system.
- Alarms for all essential functions, including cooling fan failure.
- Fully redundant power supplies
- Fully redundant control interfaces, dependent on control system.
- Design allows for interfacing to other signal types if a new requirement develops.
- Can be controlled by most current control systems:
    1. NVISION ENVY
    2. Sony
    3. Tektronix SMS-7000
    4. Philips DVS Jupiter
    5. Artel (Utah/Dynatech)
    6. ProBel
    7. AVS OMNIBUS
    8. Pesa

For very large system requirements (above 256 x 256), NVISION has also developed technology to allow cost effective expansion by using ENVOY 6256s and new secondary switch designs.

Chapter 5
Control Systems

The signal routing system is essentially the heart of a facility, supporting many if not all of the functions that are carried out to sustain operation. These include master control, production, signal acquisition, and editing. Central to this system is the interface with the operations and technical staff. While the hardware portion of the router can be fairly generic in nature, the control system must meet many conflicting needs. It must have an intuitive switching interface to support operator use. Both dedicated and software configured interfaces should be supported. The configuration utility needs to be powerful, flexible, and simple to navigate. Ideally these tools will be open platforms to allow users to customize them as required. Seamless fault resolution is needed to keep the system running regardless of the state of either the hardware or controller(s). Meeting these goals is a trade-off between feature set and cost.

Traditional Control System Topology

First Generation Routing Control Systems

Early routing control systems were limited by the technology of the time. Microprocessors were in their infancy and were expensive to use. Rudimentary serial interconnects were the only means available to link various components of the system together. Configuration utilities by necessity were embedded into the controller because PC's had yet to become widely available. Matrix hardware was built using discrete components (early systems used relays) and rarely exceeded 128 x 128 in size. Limited features were needed since a single video level and a mono audio level were all that needed to be switched. This limited feature set was actually a blessing since most selections had to made in hardware rather than with a software based utility.

The complete control system used in these first generation designs resided on a plug-in controller card typically located in the router frame. A minimal number of control panels, usually sixteen or less, could be connected via coaxial cables to the controller. An RS232 style serial interface to a dumb terminal supported what little there was to be configured. Most settings were made using dipswitches or jumper straps. System architecture was proprietary meaning only routers from the same manufacturer could be linked together and controlled. External control ports were designed for specific equipment and could not be used generically as they are today.

Sources and destinations were assigned numeric descriptions, forcing the Engineering staff to create paper legend strips for identification of names. These were difficult to manage effectively, making wiring changes arduous tasks. Early systems forced the operators to remember transcode tables (or to paste legends nearby) since source or destination selections had to be entered using numeric keypads or rotary switches. Most panels operated in an X-Y fashion by selecting a destination X and routing source Y to it.

Much was done to try and expand the functionality of these early systems. System architecture was modified to allow more sources, destinations, and levels to be supported. Source and destination names evolved to allow more than just numeric descriptions. Names could now have a two-letter/ two-digit description. Physical levels, for different signal types, were mapped with two or four character mnemonics. Control panels had alpha-numeric displays to present information to the operator, although many systems still encoded the description information into EPROM's embedded within the panel hardware. Since these early systems used proprietary hardware platforms, as new features were added the processor quickly became overburdened and rendered obsolete. System resources were limited and unable to support the continued requirement of feature enhancement. A new second-generation control system was born of necessity.

Second Generation Control Systems

Routing requirements were becoming increasingly more complex. Budgets were shrinking, and less attention was being paid to future needs. Most facilities faced the challenge of doing more with less. The rapid advance in semiconductor technology made increasingly more complex designs feasible. Microprocessor performance had increased many hundred-fold while dropping in price dramatically. This not only allowed the microprocessor to proliferate, but also allowed these same advanced processors to be incorporated into the control structure used within the router. As the feature set grew, the software running within the controller also grew in complexity.

This increase in processor horsepower allowed many new features. Much larger matrix sizes were possible, many systems supporting as large as 1024 x 1024. Larger router sizes allowed more levels, a requirement as the number of signal formats that needed to be switched grew. Broadcast and Post production facilities now had to deal with analog and digital video, stereo audio in both analog and digital forms, time code and even machine control routing.

Operators now had to support multiple functions using the same equipment. Router control panels offered advanced features intended to make this possible, such as salvo operation, multi-destination support and reconfigurable operating modes. The X-Y control panel was joined by many more variants. Since panel operation was under software control it became a simple matter to add new panel types supporting new features. The simple button per source panel could now support multiple pages and do breakaway takes. As features increased each button began to support many functions, depending upon operating mode. X-Y panels became more sophisticated in the way sources and destinations could be selected.

Two features have become part of second-generation routing control and are now heavily utilized in new facility design. The first of these is tie lines. Tie lines allow two separate levels within the router configuration to be interconnected. These levels can be local to a single controller or extend across multiple controllers in several systems. Tie lines are used primarily to support two functions; simplified interformat conversion and more economical router expansion. By locating expensive transcoding hardware within the physical tie line path, format conversion can occur automatically (as far as the operator is concerned) with source selection. Router expansion is accomplished by using tie lines to bridge between smaller routers effectively creating a larger one. While this architecture is not considered non-blocking, careful attention to the number of tie lines created will allow virtually seamless operation.

The other feature being utilized more and more is the capability of networking several router controllers together over either a LAN or WAN. This allows not only remote operation of the router from virtually anywhere a network connection is available, but also a very high level of system integration between functional areas within a facility. Separate controllers linked over a network can minimize the impact of a component failure to just a single router, keeping most of the plant up and running. Small satellite routers, purpose built for a single application, can share control panel topology with other larger routers already in use. If a satellite or fiber link is part of a tie line path then remote sites can switch feeds out to their location without local operator intervention.

As with early first generation systems, manufacturers began to enhance feature sets by tacking patches into the existing hardware and software. Mnemonics were improved, allowing eight-character prefix and suffix sets to be used for naming. More control over hardware utilization was available to deal with the ever-growing number of formats that needed to be switched. The ability to support networked control systems and tie lines added significant overhead to an already burdened processor. Solutions introduced to patch these control system issues often caused compatibility problems with previous versions. Software still ran on proprietary control hardware, although typically a PC running a GUI based program was used for configuration. The same bottlenecks that limited flexibility were returning. As end-users demanded new features to support more complex routing requirements, the second-generation control systems became overloaded. These features included determinism, logical level mapping, dynamic alias names for sources and destinations, remote control panels and routers, advanced automation, and more complex tie line management. Some of these features were driven by the fact that we have come to expect a router to do more than just route. DBS facilities switch router bus outputs directly to air, often under automation control. A post house may use router destinations to provide preview switching. As these systems grew in size, and more and more tie lines were added to make interconnections between the various systems, database management took on nightmarish proportions. Thus, the requirement for a third generation control system that married together several technologies was beginning to take shape.

Third Generation Control System Concepts

Looking back on the previously implemented control schemes reveals a common problem. Eventually the control system becomes the bottleneck that prohibits further expansion. Replacing it, much like heart surgery, is an invasive and expensive process. The data structures used to hold configuration information are typically proprietary so all of the data needs to be re-entered, often manually. Limited support for products from other vendors means existing hardware is not reusable. Operators have to relearn control system quirks since the interface will likely be totally different. NVISION's approach to control system architecture has been to support interfaces from many other vendors. They have been engineered into the products right from the start. Existing control equipment can remain in service while newer components are added. This experience has also let us evaluate the many control systems currently in use. There are systems available today that are advertised as being third generation designs. However, they are currently aimed at the larger facilities, and are overkill for basic router control.

A third generation system should be scalable and use a platform that is not proprietary. NVISION's ENVY Control System is scalable and runs on a standard platform. Basing the control system on a commonly available hardware platform will allow system upgrades to not only improve performance but to also make installing them less intrusive. Redundancy should be designed in from the start, with the level desired being user defined. Of course matrix hardware could still be vendor specific but this hardware has become increasingly reliable and failure is generally not catastrophic.

Configuration should be intuitive and the database should be open, preferably using an off the shelf database management utility. Upgrades to add features or enhancements should be simple and not affect operation. Support for third party interfaces to allow custom applications or external control should be easy to implement and reliable in operation. Features that have become entrenched in today's designs should be expanded upon. Software should allow us to use conventions that are familiar, i.e. using plain English to name system components.

Implementing this kind of architecture requires a rethinking of how various components within the plant interrelate. The intended level of control ENVY supports is completely scalable, in essence open-ended. Such features as tie line management, matrix status monitoring and embedded diagnostics are essential to allow system performance verification and fault resolution. An open platform such as that used by ENVY will allow not only generic examples of these utilities provided by the vendor, but custom designs as well. Rather than being restricted to a few limited applications, control protocols will support upload and download of complete system status.

Some of the features that third generation systems will offer include support for very large multi-stage routing and distant routing. Routers may be located hundreds of miles apart. Advanced tie line algorithms will support these interconnects and provide alternate path routing in the event the primary path is not available.

Since it is impossible to predict all the future requirements your facility will have to support, the selection of a control architecture should be made with careful consideration. The presence or lack of features today should not weigh as heavily as the ability to add them later. A system that is scalable allows for growth. Making sure the system limits will not be exceeded too quickly is a guessing game. Choosing a control architecture that is engineered to allow future expansion buys a wide margin for growth.

The Marriage with Computer Networking

Network topology designed for the enterprise market has grown in sophistication. As reliance on the communications infrastructures that support these networks increases, the same issues that are driving broadcast and post facilities to a more sophisticated control structure began also to affect the computer industry. Redundancy, fault tolerance and support for multi-stage routing are just some of the tools we use that are being implemented by computer hardware and software companies. Software and notably database creation and management tools have become extremely powerful. Software in general, along with advanced interface hardware, has made the PC a powerful data management tool. At the same time broadcast and post production facilities were facing more sophisticated control system requirements, the computer industry was implementing many highly advanced solutions. It seemed that the time was right for these two technologies to come together.

Routing control systems began a merger with PC technology in the early 1990s when second generation control systems started using PCs for configuration and download. Controllers began to sprout Ethernet ports allowing interface not only between separate controllers but also between those controllers and computer networks. It was only natural that this progressed in the following years to the PC replacing more and more of the control system. Advances in touch-screen technology in the mid-1990s made highly sophisticated PC based router control panels affordable. Several of these “home-grown” systems grew to become products on their own; with the authors starting their own companies. These include Buf Technology and Iris Technologies. But this was not the only place the PC started to make inroads. The rapid advancements in computer technology, coupled with the computer industry’s growing need to solve similar routing issues led to an overlap in technologies. This marriage forms the basis for a new third generation of control architecture. Features and scalability unheard of with current platforms will become a reality. Control systems of the past were based on manufacturer proprietary hardware for routing control. There are disadvantages to this; the time and expense of bringing a new product to market, the time and expense required in resolving problems and keeping up with changing technology and product features. Proprietary hardware is fixed in the time frame it was created and cannot easily be changed to take advantage of technological advances, having a life cycle of five to eight years. As technology advances and improves, silicon becomes faster and cheaper. Memory becomes more abundant. Processing power becomes more affordable. A third generation control system should not only break away from using proprietary hardware, but should also integrate with existing hardware technology to extend its life cycle. Many facilities have a large investment in a current routing system. A control system that supports the interface of both new and existing routers (and their control panels) to PC servers can remain technologically current and provide a bridge to the past. As PC technologies improve, the product improves, and both hardware and software advance in capabilities and speed.

Ethernet Topology

An ideal method for connecting large numbers of separate router control cards and their associated control panels would be to expand a control system using standard ethernet topology. Routers could be connected to a "router server," while control panels could be connected to a "control panel server." Each server would contain interfaces necessary to support the router hardware under it. These two servers could be connected via Ethernet to provide interoperability. System expansion is easily accomplished through additional servers linked via Ethernet. Requirements driven by the computer industry now provide for both redundant server support and ethernet connectivity, providing a high level of fault tolerance. And Ethernet hardware is also inexpensive and readily available!

Third generation control architecture should also be expandable so as to appeal to the majority of facilities. Servers interconnected with Ethernet meet this requirement. See figure 5-1. Initial systems would consist of a single control system server, supporting both control panel and router connections. This system topology would be attractive to a small television station or post production facility with limited routing requirements. Router sizes, number of levels, and number of control panels would be limited. Flexibility in the software would allow additional features to be added easily. The architecture installed is identical to that needed to support larger systems so expansion will not be difficult or intrusive to daily operations. An infrastructure can be put into place that does not need to be completely revamped each time a new requirement comes along.

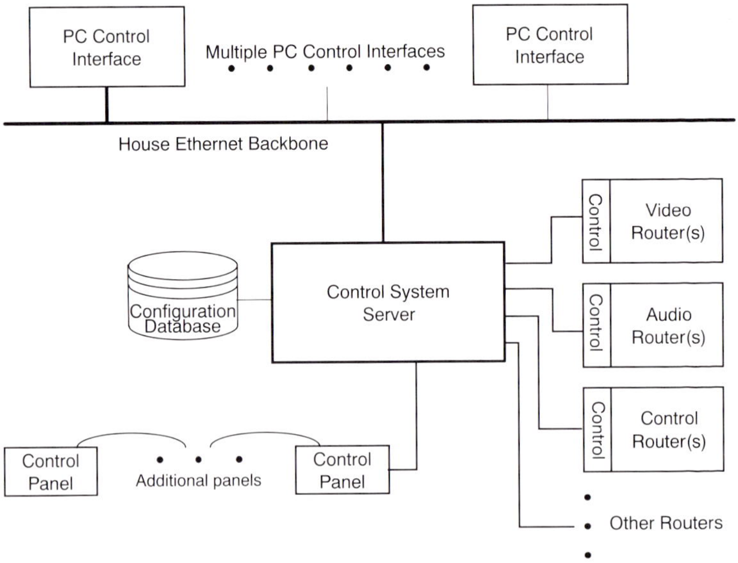

Figure 5-1. Single Server Control System

The simple system is expanded by utilizing additional control modules. See figure 5-2. As software support for third party equipment becomes available additional devices could be added to increase flexibility. The facility may elect to segregate routers and control panels using separate control modules if increased fault tolerance is required. Unlike the systems in use today these changes could not only increase system features, but also performance. Adding a second control server for example may also mean an increase in processor horsepower simply because the price point for this hardware has dropped. The original, slower system becomes the backup.

Note that in these and the following examples the connection between the control system and the matrix hardware is done with a high-speed serial interconnect. Connections with other system devices may use any one of a number of available interfaces available for the PC platform.

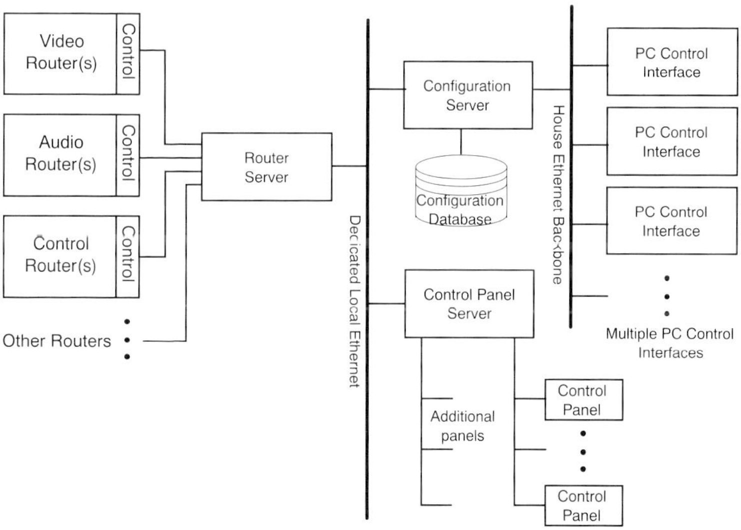

Figure 5-2. Multiple System Control System

Control panels and matrix hardware could be intermixed on the same control system server, with multiple systems supporting different areas of the facility (figure 5-3). Ideally, the distance between these servers should not matter, as long as the distance is supported in the interconnection methodology used between modules. Since this interconnect is not proprietary in nature, it needs only be sized to support the intended application. For example, Ethernet connections using twisted pair cabling may be used within the main equipment room to interconnect matrix hardware and other system devices. Thin net could be used from the server to control panels due to its ease of providing a daisy-chained path. Fiber might make sense for a high-speed interconnect between two servers located some distance apart. If only a LAN needs to be implemented, then gateway hardware out to a WAN is not needed. This is one of the benefits to using Ethernet. Support for both LAN and WAN architectures is part of the interconnect specification. Network Interface Modules (NICs) are available that support 10 mbit/sec or 100 mbit/sec connectivity. Adding more functionality involves simply adding appropriate hardware onto the network and the software to drive it.

By utilizing high-speed NICs and wide-area networking (WAN), portions of the control system could be located many miles away (figuren5-4). In an ideal environment, the manufacturer could allow completely independent control systems to be linked via tie lines and WAN technology. The configuration of such a complex system becomes extremely involved, as both systems must include mnemonics from the remote system's routers and must be kept up-to-date independently. The system administrator(s) of such a complicated system must take care in the naming of the sources and destinations to ensure all users know exactly what sources are being selected and from which system. It must be noted that by the nature of network technology, control systems connected through a WAN are unlikely to be deterministic, unless time coded switch commands are used.

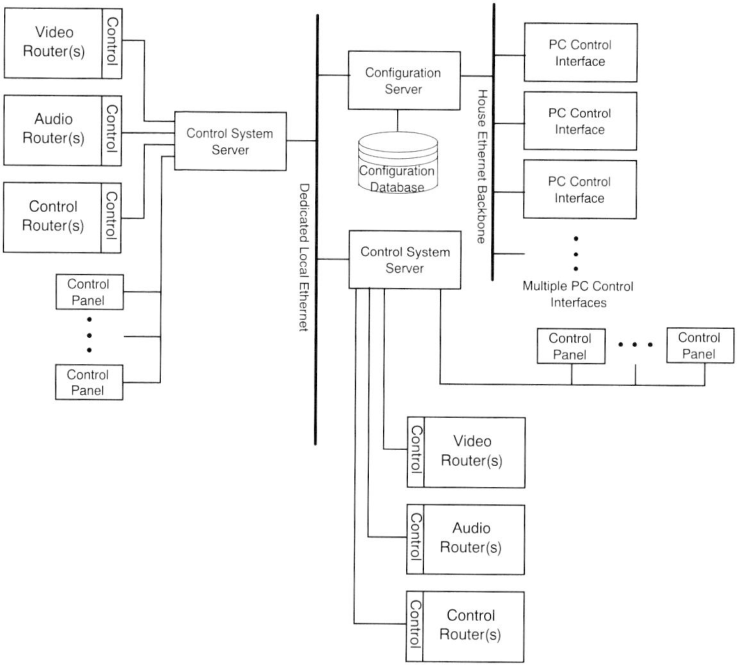

Figure 5-3. Multiple Server Control System

In this era of downsizing and the combining of facilities, a control system that supports a WAN-based topology becomes more and more important.

Redundancy

A routing control system must offer redundancy as an option. Redundancy can take on many different forms but ideally should be capable of being tailored to a specific need. Most important is router control redundancy. If the control of the router fails then operation ceases. Other system components such as control panels and matrix hardware should support redundancy but often cost constraints limit the importance of this need as there are often other control panels in the facility and spare router cards to fall back on.

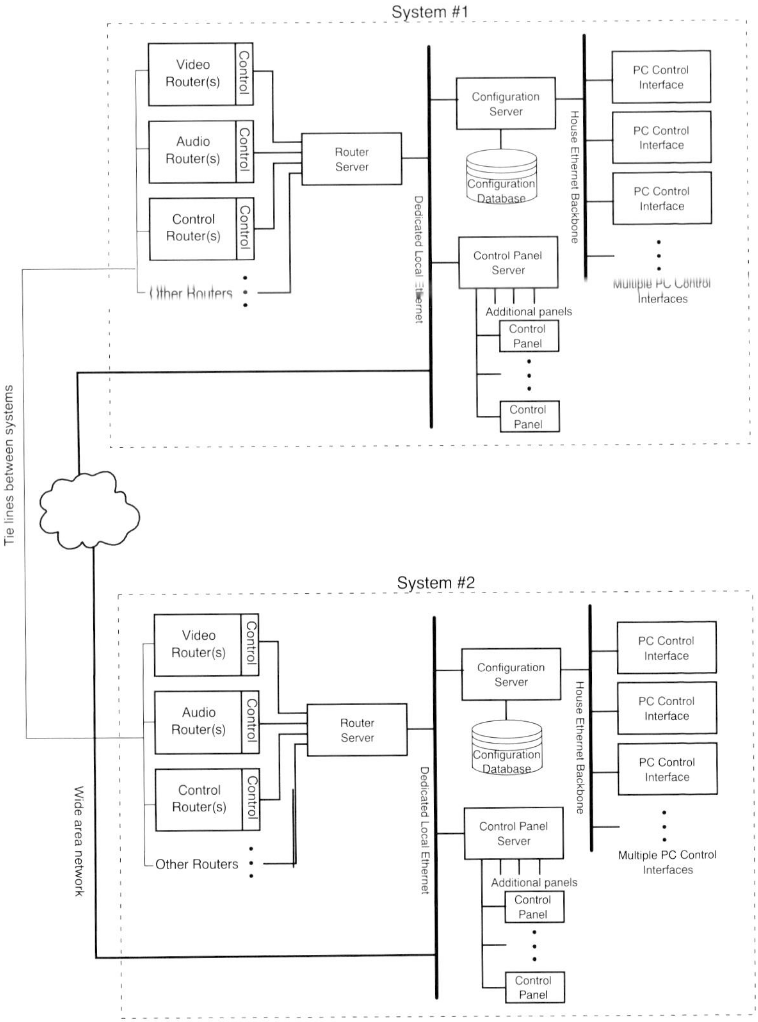

Figure 5-4. Wide Area Network (WAN) Control System

A third generation control system that utilizes standard PC technology could use this same technology for redundancy. There are several different methods to accomplish this. A simple method is to use a PC with redundant, hot-swappable power supplies. If the power supply in the PC fails, the computer will switch over to the redundant supply. The user can then change out the bad supply.

Fault-tolerance has become more widely available in the computer industry, thereby providing another simple method for redundancy. This method would use PC servers with hot-swappable power supplies and hot-swappable disk drives with Redundant Array of Independent Disks (RAID) disk mirroring. Several server manufacturers also provide redundant NICs and processors. Using two PC servers and connecting the routers and control panels in parallel to both can enhance redundancy.

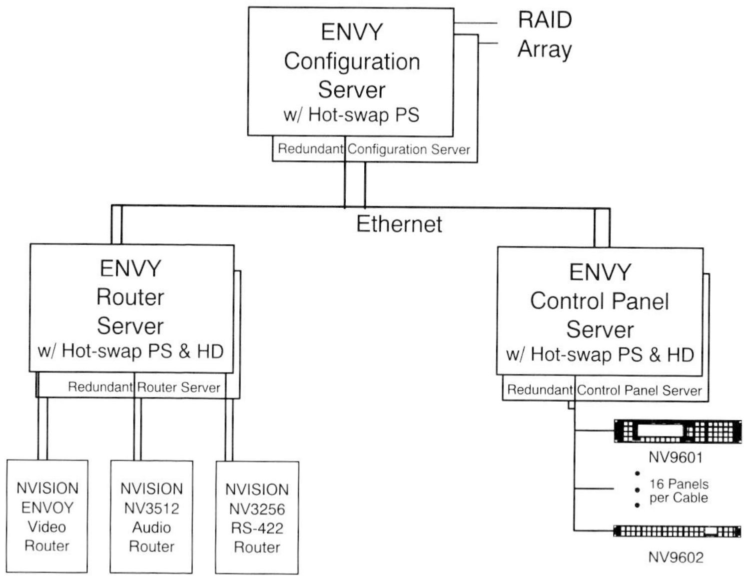

Figure 5-5. Redundant System

Yet another way to accomplish redundancy is to utilize Microsoft's NT Enterprise Edition's™ Cluster Server software. See figure 5-5. Clustering is defined as a group of independent servers working together and allows one server to automatically fail over to another server. It is common for the clusters to share an external RAID array with disk mirroring and disk striping with parity to ensure configuration and status availability. Both cluster servers can provide computing power at the same time, providing load-balancing capability. Resources such as routers may be connected to ports on the servers in the cluster. The control port on the main router control card is connected to a port on one server and the control port on the redundant card is connected to another server. If one server fails, the other server will take over control of the router. ENVY can take advantage of all these forms of redundancy.

Determinism

The need for determinism was brought up earlier. A predicable latency between the request for a switch being made and when that switch actually occurs has always been a desirable feature. Surprisingly many current routing systems do not offer it! With routers finding greater application in the on-air chain of a broadcast facility this feature is no longer considered just desirable, but a necessity. Any third generation control system must support deterministic switching. The following list defines the different types of determinism as it relates to video switching.

1. All levels switch at the exact same time (time code aligned).
2. All levels switch within the same vertical interval (VI aligned).
3. All levels switch at a time determined by the control system (time frame aligned).

A time code aligned switch is VI aligned (as long as the time code is locked to house reference), but a VI aligned switch may not be based on time code. A control system should be capable of time code aligned switching provided the command is received at some defined time in advance of the desired switch point. The amount of time required for the command to precede the switch will depend on many factors, i.e. the speed of the control system, the number of router control modules, the number of routers, the number of levels in the switch, etc. This time, as long as it is constant, should not affect operation. The switch will be deterministic regardless of which control servers the routers are connected to. Providing deterministic switching when Ethernet is used as a communications medium is not a simple task. Ethernet by nature is not deterministic. When a message is introduced into the network the only guarantee (and not a real certain one at that!) is that the message will arrive at the specified destination. When it will arrive is not implied or specified. Care should be exercised when modifying or interconnecting networks. One of the reasons a separate control path is used for matrix communication is to allow tight control over what traffic uses this medium. Attempting to use an existing facility LAN for this path will almost certainly result in switching errors. Similarly an external router control program running on a PC connected to the router via a gateway or other network bridge would not be deterministic. When the server receives a take request, the switch time becomes a known entity. The lack of determinism stems from the fact that getting the take request into the server is not predictable. Normally control panels that require operators to push a button to initiate a switch are inherently non-deterministic. The operator does not care exactly when the switch occurs, only that it happens soon. Automation systems on the other hand must know the latency. This is one carry over from early control schemes. Interfaces to external equipment still need some definition that may be brand specific simply to allow specifications to be guaranteed. This may entail some custom software or a specified port the external device must use, quite likely even both. NVISION's ENVY third generation control system specification has full support for both native NVISION equipment as well as a wide range of third party gear. There is one caveat however, the system is only capable of providing this capability if the equipment itself supports such a feature. Switching alignment may not be available when controlling other manufacturers' routers. Also, determinism is only possible for routers on the same control server unless time code based switching is used.

Open Architectures

The SMPTE Recommended Practice for Routing Switcher Type-Specific Messages for Remote Control of Broadcast Equipment RP 191 (ES-Bus Switcher Dialect) has become a common means of controlling and statusing routers from external control systems. For example, several current production switchers, audio mixers, and under-monitor display systems use this protocol to gain entrance into other vendors routing systems. The list of manufacturers adhering to and providing this protocol support is growing constantly. The list of products available for control via this interface is also increasing. Any new control system should have sufficient handles built in to allow easy expansion of what can be controlled and how. This protocol document is available from SMPTE. If large-scale automation or a high level of integration between the router and other system devices were planned, study of this document would be prudent.

The Philips DVS implementation of this protocol is the most widely accepted version. There are several differences between this implementation and the recommended practice. For example, the SMPTE document requires polling and select addresses; Philips' document states that "No polling or select addresses are necessary or accepted." With this implementation most of the common messages, as well as the time line and procedural message structures, and several routing switcher type-specific messages are not supported. The Philips implementation also does not support the concept of multiple matrices. To provide flexibility as much of the SMPTE specification should be supported as practically possible.

The NVISION device control protocol incorporated into the ENVY control system will support as much of the current implementation as practical. Since it is a published protocol, intended to be embedded into NVISION's new ENVY control system as well as the control systems of other manufacturers' who wish to control NVISION routing switchers, compatibility is essential. Using this protocol, newer NVISION routers may be capable of time code aligned switching. This protocol is also used to control NVISION delay and mixer modules.

The NVISION Application Programming Interface (API) is a new public interface that will allow other systems to control routers within the ENVY control system. It is an interface that many production switchers, audio mixers, under-monitor display systems, and automation systems may choose to use. One of the issues faced when implementing new control architecture, especially one of the magnitude that ENVY represents, is defining and then gaining support for new command protocols. By defining a feature set that allows flexible interface options that many users desire, other vendors will find it advantageous to include support to satisfy customer demands. A third generation control system needs to have this forethought as part of the design specification, breaking away from traditional closed architectures. This protocol document is available for the asking from NVISION.

Automation systems

SMPTE RP-191 protocol mentioned earlier, has become a common method for automation systems to control routers slaved to control systems. It provides the automation system a method of deterministic switching based on time code, provided the rules defined earlier are followed. Many vendors currently use all or part of this specification.

Master-21 Protocol, published by Tektronix, is a common method for automation systems to connect to master control switchers, The facility may or may not incorporate a router or facility control system. As defined, this protocol does not provide for deterministic switching of the associated router. M21 protocol has been around for many years and was one of the first true interfaces into the on-air chain. It should be considered a first generation product that has been resilient enough to survive many patches throughout its life. Recently when the M-2100 was released it was upgraded to include additional features. Support for this protocol is essential due simply to the large number of systems currently in use.

The NVISION API, as hinted at earlier, is a newly defined method for external devices to control router hardware that is slaved to a control system. Being a third generation implementation it addresses and provides solutions for many of the issues plaguing today's interface specifications. It provides an automation system with time code-based determinism as a start. However the intent of this new API is to go beyond just the interface to automation, and to also support more advanced facility control architectures where many devices become part of the network. The open structure and freedom from proprietary control hardware will hopefully allow a growing number of devices to reside and become part of the overall system. The API is not however restricted to be being the system housekeeper...it can also become part of a larger structure under control from an external engine such as Omnibus. The command set will support deterministic control. Switches from an external interface to the ENVY control system using the NVISION API (or SMPTE RP-191) should be either time code or VI aligned, depending on the command set used.

Master control

Early master control switchers were little more than video switchers with audio control slaved to them. It was common to use two router busses as program and preset. This early topology started a long standing marriage between master control and the facility router. The interface between the master control switcher and the router control system takes many forms. It may be as little as mounting a router control panel into the switcher frame. More typical is a linked arrangement where the router provides input expansion for the MC switcher. The MC switcher may or may not be slaved to the router control system to achieve this link. Usually a subset of the router source table is made available to the MC switcher through configuration, and router mnemonics are displayed at the MC switcher to allow source identification. Most master control switchers currently provide the mixing and switching functions and typically support up to 16 inputs; preset, program, clean feed, bypass, and auxiliary busses are the usual outputs. If used as a stand-alone unit, all sources to be switched to these busses come from these dedicated inputs. With the current control architecture the interface to the router, as well as other devices such as automation control, is done serially. More recent systems support Ethernet interfaces for these functions. A third generation control system should support these devices as nodes on a network. This allows simple integration between the traffic system, commercial playback, on-air switching and the router. Router control points can be embedded into the MC switcher's control panel and be made to operate similarly. A single network connection can then support multiple command streams out to various devices. With the router linked into the MC switcher, preset source information can be stripped from the daily traffic log and used by the router control system to make sure required sources are available for playback to air. Or, more simply these source selections can be commanded from an external control system. The importance of an open architecture such as that used by ENVY, with an Ethernet-based communication medium, starts to make sense. The MC switcher is inherently deterministic by design. The interface to a router control system may be deterministic if the control system it is slaved to provides deterministic switching.

Future Possibilities

Scheduling/billing system interface

Many post production facilities utilize scheduling and billing programs specific to their industry. Broadcast operations need on-air reconciliation to validate spot playback. Marrying these systems with the router control system is a natural progression. The scheduling program uses the API to send switch commands to the control system, based on a program or scheduling log. When a client arrives, the edit room is configured with all the desired equipment. To-air logs are printed from the router in addition to the MC switcher.

Implementing this could take several approaches. The control system could support switching by configured device classes. For example, the scheduling program might ask the control system for a D1 class device. The control system would assign a D1 class device by switching it to the desired destination. Other devices required would be switched similarly. Destinations could be grouped by functional area. The control system then reports back to the scheduling system the actual devices assigned and any conflicts encountered. Billing systems could monitor which devices were acquired and their utilization through the router API, calculating charges accordingly.

Router control

Advances in network speed will provide a future method for matrix hardware to reside directly on the facility network as a node. High-speed connections at either 10 or 100 mbit/sec will replace slower connections, commonly 38.4 kbps, 115 kbps, or 1 mbit/s. There are many advantages to this...matrix hardware can be located where cabling is most convenient, temporary routing may be set up on location and linked to the facility via the Internet, remote control points can be set up anywhere they are needed. Using the Internet may preclude any determinism, but as speeds increase, the control system may be able to provide near deterministic switches. The use of time coded switch messages will provide determinism, even when using the Internet to switch the router.

Client tasks

When an operator in a post production facility performs a client task or job, he or she performs specific functions that use resources pulled through the control system. These functions include, but are not limited to making routes, locking or protecting sources and destinations, using tie lines, and using control router resources. A control system could manage these tasks for the operator based on a task ID number. All control system resource usage could be attached to that task number. Switcher settings, machine control, and clips located on a video server could all be uploaded and made part of the link. When the operator completes the task, the control system could automate freeing all the resources associated with that task. This could include releasing locks, protects, tie lines, and data routes. Client tasks could also be integrated with scheduling and billing systems. Duplication could be handled automatically, even supporting product delivery via the Internet or other delivery means.

Chapter 6
Embedded Audio

The SMPTE 272M standard specifies the method of formatting a maximum of 16 audio channels in the ancillary data space of the SMPTE 259M serial digital video standard. Further, SMPTE 299M also specifies the architecture of up to 16 audio channels that can be incorporated within the SMPTE 292M standard for the high definition television serial digital interface.

Embedding audio within the video signal offers advantages over traditional methods of running separate audio/video systems, particularly in broadcast facilities where limited audio breakaway is required. Utilizing embedded audio offers some attractive benefits: simplified system design, reduced cable requirements and DA count, a single routing system, and excellent cost savings.

In the past there have been problems with embedding and disembedding. One of these is switching errors. When a switch is made between two video sources that contain embedded audio data, it is difficult to resolve a clean audio transition at the receiving end. Figure 6-1 shows the timing relationships between PAL, NTSC, and AES. At 48 kHz, there are five AES blocks during each PAL video field and 4.170833 blocks for an NTSC field. If a video signal is used as a genlock source for AES signals, the frame alignment and phase relationship between audio signals is arbitrary. In this circumstance, regardless of whether a signal is embedded or not, clean audio transitions are difficult to achieve.

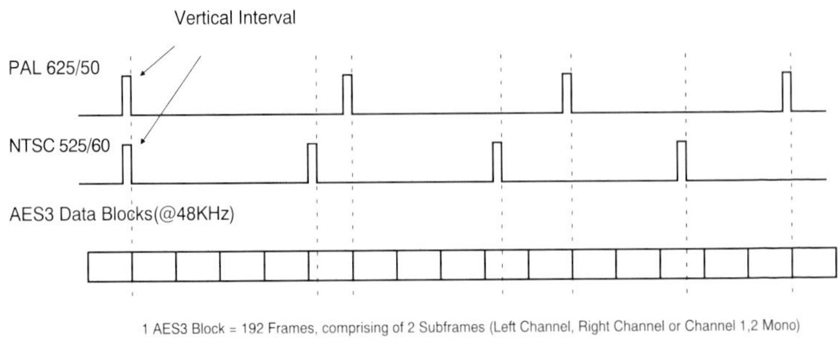

Figure 6-1. PAL, NTSC, and AES3 timing relationship

Some manufacturers have provided solutions to the switch error problem by including audio sample rate converters in the embedder and disembedder designs. This has the potential of minimizing the problem by re-sampling and re-framing the audio to a common reference, but removes any possibility of maintaining the original audio phase.

Then there are the multi-channel difficulties. When more than four channels are required, the normal technique has been to cascade embedders and disembedders. See figure 6-2. Cascading is expensive and relies on the ability of the embedder to determine if current ancillary data content exists and where and how to allocate new data. The more channels inserted, the more difficult it becomes to determine channel location at the receiving end. Plus, with the advent of surround sound in general usage, phase alignment becomes all but impossible with cascading. Multi-channel embedding and disembedding creates indeterminate phasing across channel groups, possibly degrading the effect of surround sound.

NVISION has developed innovative embedder and disembedder modules for 270/360 Mbit SDI signals that solve these problems. (HD-SDI will follow as silicon is developed.) All audio inputs are accurately timed to the house AES reference (which is locked to the master video clock, see figure 6-3), to create error-free switching if the NTSC

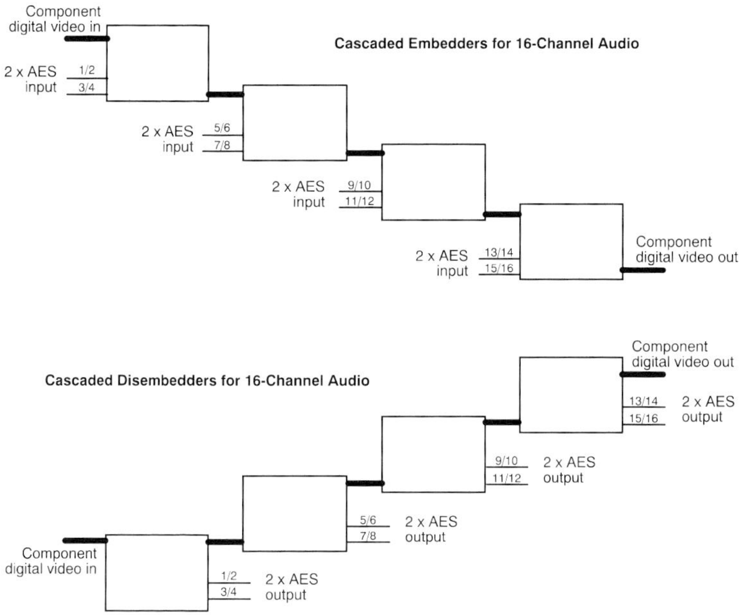

Figure 6-2. Cascading embedders and disembedders for multichannel audio

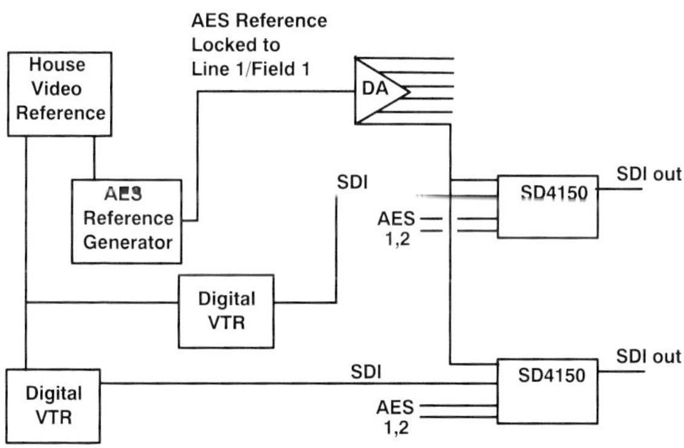

Figures 6-3. Embedders locked to house AES reference

video path lengths are the same. (PAL is not a problem.) Should NTSC path lengths differ, or the audio be embedded by products other than NVISION, then the re-framing (re-timing) ASIC in the disembedder comes to the rescue. If the AES framing is disturbed by a video switch, this ASIC will continue to provide a constant AES output, thus eliminating the possibilities of receivers losing lock and requiring a finite (and audible) recovery period. See figure 6-4.

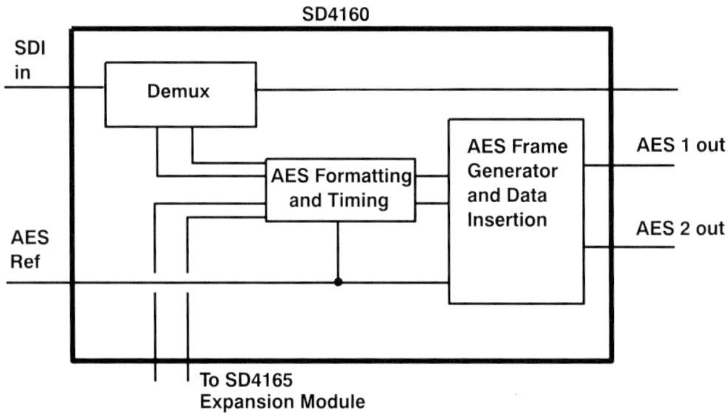

Figures 6-4. SD4160 disembedder block diagram

With these new designs, cascading of modules is no longer required to support multi-channel expansion. NVISION's expansion modules can add an additional 12 audio channels (for a total of 16) to these novel embedding and disembedding products. See figure 6-5. Not only does this save money but it also solves the previous problems of identifying channel location when using cascaded modules.

All data allocation difficulties are resolved by selectable group assignment (fig. 6-6). Further, the proprietary AES re-framing ASIC ensures that all inputs are accurately sample aligned across all four groups. Transport for uncompressed multi-channel audio and surround sound mixes is now reliable with total phase control!

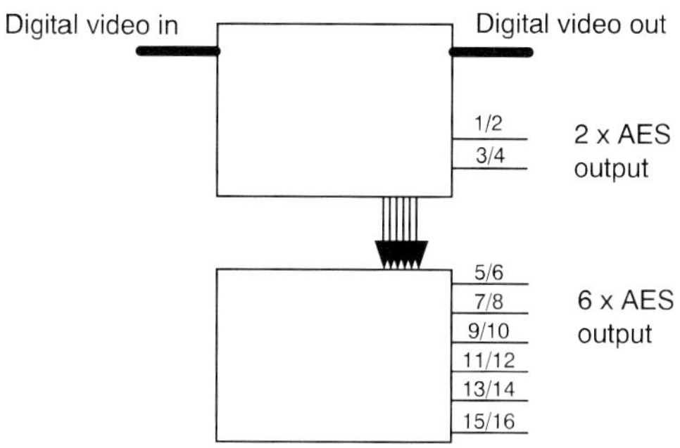

Figure 6-5. NVISION multichannel disembedding

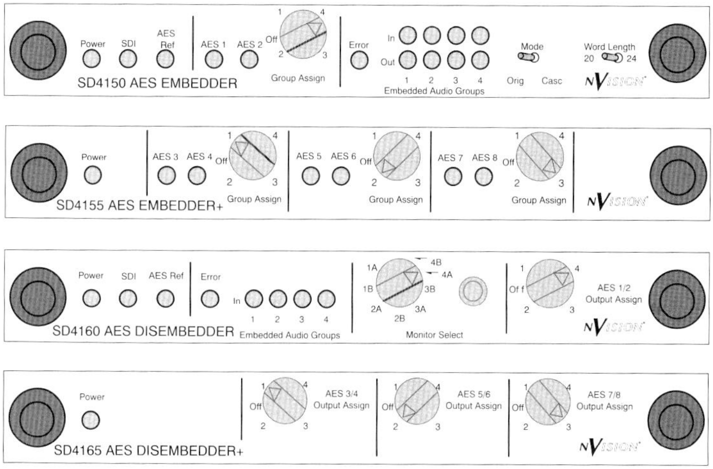

Figure 6-6. SD4150, SD4155, SD4160, SD4165 front panels

There are still many things that need to be considered when designing a system that utilizes audio embedding. Once audio and video are combined, audio and/or video insertion and mixing may no longer be possible without first disembedding the audio. Even inserting a station logo on video via a master control switcher could disturb the embedded audio data, therefore limiting your ability to manipulate any signal without first routing it to the appropriate disembedding and re-embedding devices.

Another point of consideration is that any video equipment in the signal path might introduce possible delays. As audio data is embedded in almost every line of video, realignment of video data can be detrimental to the audio information. It may still be recoverable, but it will certainly not be synchronous with other signals and it will no longer be possible to provide error-free switching with another signal that did not take the exact same path. See figure 6-7.

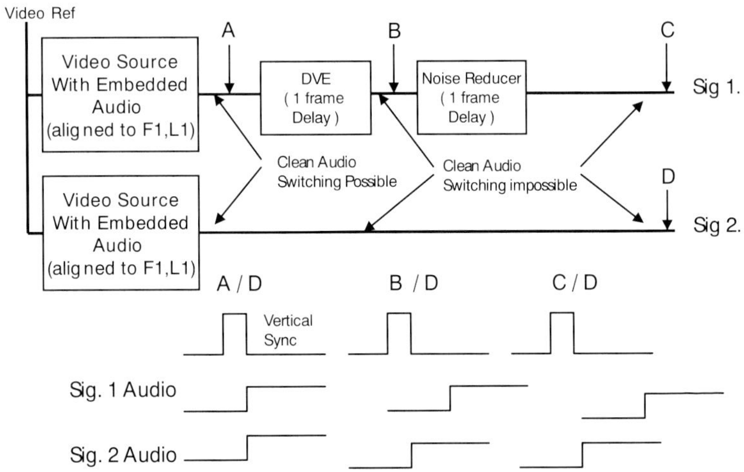

Figure 6-7. NTSC or 59.94/29.97 systems, audio/video timing relationships

Embedding is common today for distribution to remote receive and retransmission sites, but input timing can cause some very undesirable problems.

Where video inputs need to be re-synchronized to a local reference, decisions must be made as to how and when to extract the audio. If the audio is extracted after the video frame synchronizer, any frame drops or repeats necessary to retime the video will result in the deletion or repetition of a very large number of audio samples, with objectionable effect. However, the digital audio output of the disembedder will be correctly clock locked to house timing (fig. 6-8).

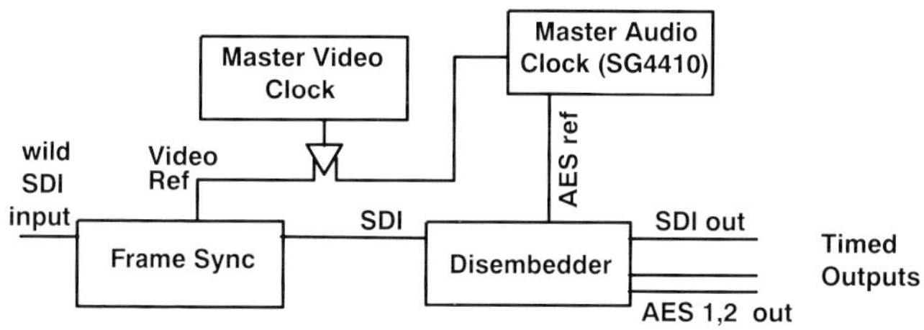

Figure 6-8. Disembedding after the frame synchronizer

Alternatively, the disembedder could be placed ahead of the frame sync to recover the audio before any samples are destroyed or repeated. In this case, the audio output will be clean, but still 'wild' in comparison to house reference. In order to retime this signal it is necessary to include an audio sample rate converter (SRC) in the signal path. (For this reason some manufacturers include SRCs within the disembedder design). See figure 6-9.

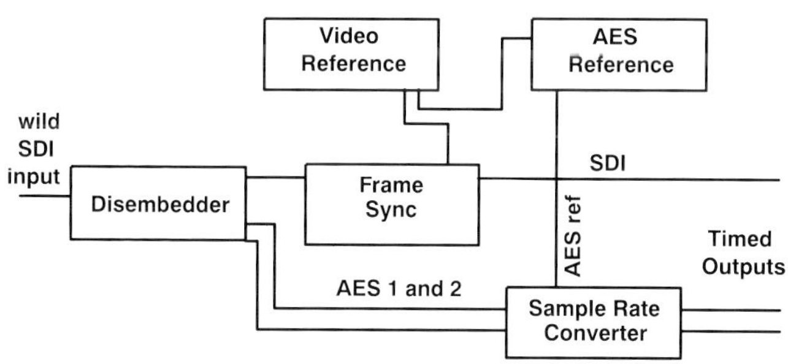

Figure 6-9. Disembedding prior to the frame sync

This second method would seem to be preferable, with the only downside being the requirement to add audio delays to compensate for the frame synchronizer. However, if there is a need for multichannel audio, the SRCs will almost certainly generate sample slips between AES streams, causing the loss of inter-channel phase coherence. At NVISION we felt it necessary to find a method of synchronizing the incoming audio to house, without the need for SRCs. The solution is somewhat unconventional, but in many cases totally viable.

By accurate control of the audio output timing and careful control of the disembedder buffers, we have developed a method of dropping or repeating single AES samples during an audio synchronization process. The NVISION disembedder (SD4160) is the equivalent of the video frame synchronizer; it must be fed with a house AES reference to which the output audio framing is locked, see figure 6-10.

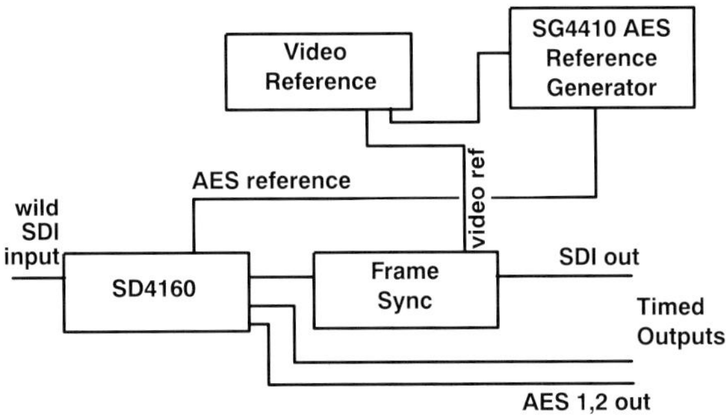

Figure 6-10. Using house audio reference with SD4160

The wild audio input, extracted from the video signal, will be then re-timed by dropping or adding samples as the incoming signal over/under-runs the local buffer due to local and remote clock time differences. Dropping or repeating an audio sample is not transparent; the frequency at which this occurs will determine its acceptability. Any sample drop or repeat will have an adverse effect on the signals THD (Total Harmonic Distortion); as the frequency increases the THD will worsen. In listening tests, we have found that with clock differences (local and remote) of up to plus or minus 10Hz of sub-carrier frequencies (most video generators have a tolerance of plus or minus 1Hz) are generally inaudible.

By using this approach, the SRCs can be eliminated. Thus reducing costs, but more importantly, it maintains audio phase. Even though samples will be dropped or repeated within the SD4160, it will be a synchronous event. Therefore if multichannel audio is embedded via an NVISION SD4150/SD4155 combination and disembedded via an SD4160/SD4165 locked to local house ref, your surround sound mix will still be totally phase accurate, although the THD may have suffered a little.

Of course, you can still use these products in a traditional fashion, but we would recommend that you test this approach before making a decision.

Compressed multichannel audio data can also be embedded (Dolby Digital, Dolby E, MPEG, DTS etc.), provided that it conforms to AES3 architecture. However, none the application examples above can be utilized to disembed these signals.

As the compressed data will be distributed across several AES frames, drop or repeats will prevent the signal from being successfully decoded without elaborate forward error correction. If sample rate conversion is used, the input signal is re-sampled and will no longer be an exact copy of the original data, again eliminating the possibility of successful decoding.

There are three alternate ways to manage these signals:

- Disembed the data from the wild feed and make no effort to synchronize it to house. This is okay if the signal is to be passed through without edits; however, video path delays will make it difficult to maintain lip sync.

- Disembed the data and record video and audio signals directly. Subsequently, the recording can be played synchronously.
- Disembed the data from the wild feed and immediately decompress it. The baseband signals can then be sample rate converted to house time and distributed in the normal manner. However, you will now have a minimum of six audio channels (three AES streams) to manage and sample rate conversion may have created phase slips between the AES signals, possibly degrading a surround sound mix.

Embedded audio has its place; it can be a valuable cost saver in new systems and it's a great way to distribute multi-channel. Serious consideration should be placed on how to utilize the new features and functionality provided by embedders and disembedders during system design. Any product that is under consideration should be thoroughly tested for the specific application. The errors that can occur with some products are often very subtle. So, if you decide to embed, choose carefully.

Chapter 7
Fiber Optic Signal Distribution

Introduction

Fiber optic technology has long been a staple of the telecommunications industry because of the need for a long distance, high-bandwidth communications medium. With the ever increasing data rates found in the typical broadcast or post facility today, especially with the addition of HD, fiber optic transceivers can simplify inter and intra plant routing.

Today's serial interconnect technology allows SD bit rate signals to travel approximately 300m before equalization and/or reclocking are required. HD rate signals face even tighter constraints. Distances beyond 100m are challenging, and require careful attention to cable choice, connector installation and number of reclocks, to mention a few design issues.

Fiber Optic Transceivers permit much greater distances between source and destination, allowing simplified distribution of signals between floors and functional areas within the facility. Modern Fiber technology is highly reliable, offering much longer MTBF on the laser diode, increased receiver sensitivity and greatly simplified connector/splice installation. Some other benefits obtained by using fiber include immunity from EMI/RFI, elimination of ground loops, low cable weight, reduced cost per meter, and a much smaller physical size when compared to coax.

Applications for SDI / HD-SDI

Fiber is ideally suited to run high bit rate signals such as 1.485Gbs HD over long distances. Some typical studio applications include...

Simplified signal distribution over long distances
If cable runs exceed 100m fiber begins to make sense, and in fact may be the only option! Using conventional coax-based distribution requires expensive reclocking DAs and associated trays scattered throughout the plant.Successive re-clocking also adds jitter, eventually rendering the signal useless. Fiber suffers from none of these anomalies. The output of the fiber transceiver is as clean as the input. If embedded audio is used, up to 16 channels of audio as well as the HD video signal can be sent over distances of up to several kms.

Master Control to Transmitter
Fiber can be used to carry multiple video and audio feeds to the transmitter site. A single fiber optic cable will usually contain from 2 to 30 or more fibers. These can be used to carry both SD and HD signals economically to the transmitter. Using fiber also provides a measure of protection against a lightning strike to the tower, which would cause even more significant damage to the facility. While pulling fiber may initially seem expensive, the lower cost of optical cable and its long service life can outweigh the higher labor costs involved during installation.

Equipment interconnects
Many video servers and graphics workstations use fiber as an I/O connection. Integrating fiber into your facility allows you to take advantage of these technologies.

Point-to-point video delivery
If you are running out of room, fiber allows easy expansion into additional buildings or nearby space and is more robust than microwave links. Fiber can be used to

interconnect two facilities located across the street or across town with an economical, highly secure path. Fiber cannot easily be tapped into or received by unauthorized persons and so is ideally suited to provide the link between facilities where security is a concern.

Making Fiber Work

Types of Fiber

Fiber optic cable looks very similar to traditional coaxial cables from the outside. In principal it operates as a waveguide, channeling light rays from one end to the other. Light entering the glass core is held inside by refraction off of the boundary between the core and cladding. A typical optical cable consists of this glass core (the fiber itself and the surrounding cladding), a protective encapsulation layer and finally an external jacket that provides strength and resistance to abrasion. Figure 1 below shows this construction. Since this package is still quite small physically it is common for several fibers to be bundled together within a single cable, along with a strength member. A fiber cable should not be pulled haphazardly, since the glass core is fragile and if improperly handled can be stretched or otherwise damaged. Depending upon the manufacturer, the cable will either have a high strength outer jacket or an internal strength member, or possibly both. The strength member is typically fiberglass, kevlar or even a steel wire. During installation the cable should always be pulled by either of these components, following the manufacturers recommendations. Failure to handle the cable properly during installation will lead to problems later; many of these problems may be difficult and expensive to remedy.

Optical cable is available in two flavors, Single Mode and Multi-mode. Mode refers to how many paths (or modes) the light ray can take as it travels through the cable. Single mode fibers are always step index; multi-mode can be pur-

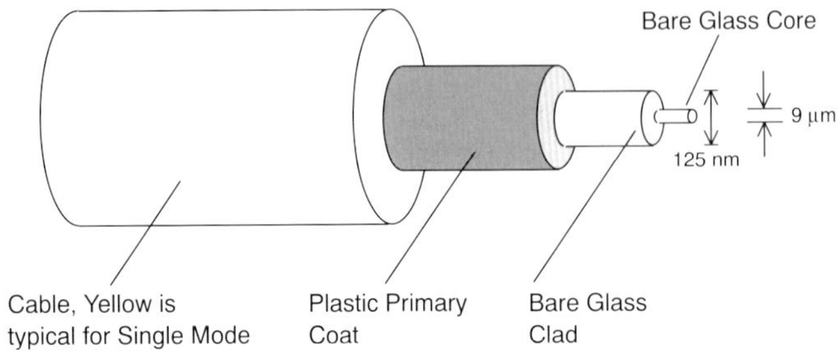

Figure 7-1. Typical fiber construction

chased as either step index or graded index. These terms refer to how light is carried down the cable. A step index design relies on a sharp difference between the refractive index of the core and cladding to keep the light energy trapped inside the fiber. Graded index cables use many different layers of glass to gradually bend light rays back towards the center (fig.7-2b). Single mode fiber allows only one path for the light ray, tightly controlling reflections from the boundary between the core and cladding (fig. 7-2c). To minimize these reflections, the glass fiber is kept small in diameter. In this way only one transmission mode is supported and light energy exits the fiber with a minimum of modal dispersion. A multi-mode design allows the ray to follow many possible paths as it progresses down the cable (fig. 7-2a). This introduces distortions because the light rays do not all arrive at the destination coincident with each other. The result is reduced bandwidth and transmission distance.

There are several specifications used to characterize optical cable. These are outlined below.

Maximum Attenuation, specified in dB/km. Optical cable attenuates the light energy as it travels through the cable. The level of attenuation is the sum of all of the losses induced. For single mode cables operating at approximately 1300 nm this number will be around 0.35 dB/km.

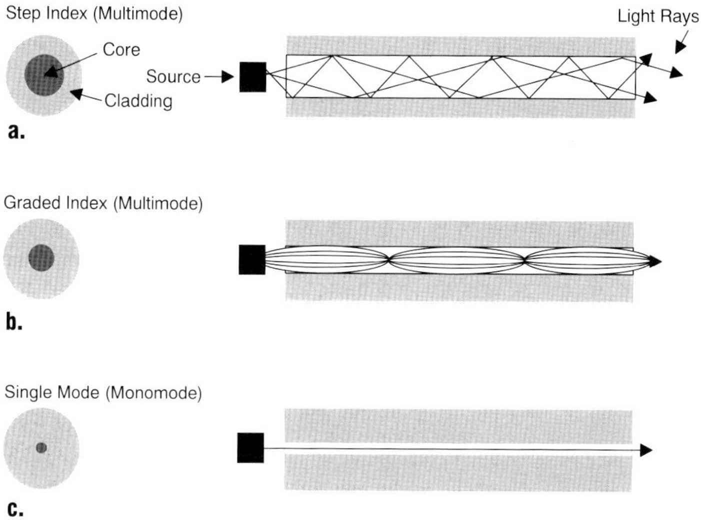

Figure 7-2. Transmission Modes

Numerical Aperture, no units. A measure of the fiber's ability to collect optical power from wide input angles. Larger values imply greater coupling efficiencies with large area sources such as LEDs. Typical values range from 0.1 to 0.4.

Bandwidth, specified in MHz-km. An operating wavelength must be specified for this to be a meaningful specification. This number states the typical available bandwidth of the fiber at a distance of 1 km at the specified wavelength. Typical values for single mode fibers are 500 to 1,000,000 or more MHz-km.

Maximum Dispersion, ps/nm-km. This is a measure of the difference in arrival times of varying wavelengths. For single mode cable used at the distances found in a typical facility this specification is not really relevant.

Minimum bend radius, inches or mm. Vendor specified, specific to each cable type. Either the absolute minimum, or optionally the allowable range of diameters, that the fiber can be bent to. Care must be taken during installation to not operate cables anywhere near this specification. Bends in the cable can lead to an increase in attenuation.

Single mode is recommended for video applications as it offers greater bandwidths over longer distances. It's lower losses also allow for some errors in installation of connectors and splices, although as with any transmission medium care should be exercised to make sure losses are minimized. With the very small core diameters used by single mode fibers, connector and splice installation can be challenging. Three operating wavelengths, or windows are available (fig. 7-3). All are in the infrared region of the spectrum. To a certain extent all single mode fiber cables will operate at any of the available wavelengths, although because of the lower losses 1310 nm is often the one used.

Fiber can be purchased that is optimized to operate at several wavelengths simultaneously. This is advantageous if higher data rates or a bi-directional path are needed. Using wave division multiplexing, two lasers drive the fiber at differing wavelengths. Optical filters at the receiver direct the incoming light to the appropriate detector. SMPTE 297 outlines fiber when used for SD rate signals. SMPTE 292 specifies the recommended interface practice for HD serial/fiber interconnection and is worth studying if HD in any form is going to be integrated into your facility.

Other factors make single mode operating at 1310 nm desirable. One is lower dispersion loss. Dispersion losses take on two forms, modal and chromatic. Modal dispersion (figs. 7-2a, 7-4a) occurs when the light rays each take a different path through the fiber. At the receiver each light ray arrives at a slightly different relationship spatially, broadening the width of the light impulses and effectively lowering available bandwidth. This effect adds the equivalent of jitter to the signal. Chromatic dispersion (fig. 4b) occurs because each discrete color of light travels through the fi-

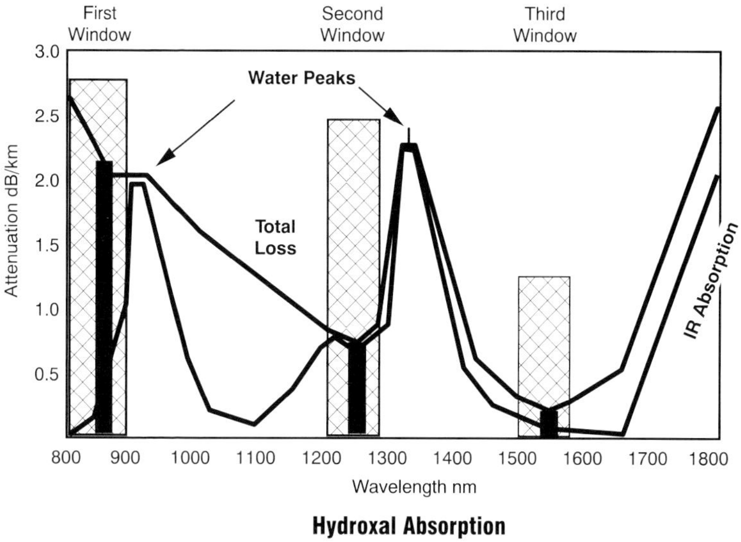

Figure 7-3. Operating wavelengths

ber at a slightly different speed. This is analogous to propagation delays in coaxial cable. Again, the result of this effect at the receiver is a broadening of recovered pulses. Single mode fiber only suffers from chromatic dispersion, unlike multi-mode fiber which is plagued by both forms. Making sure the output of the optical transmitter is spectrally clean can minimize chromatic dispersion loss. Common laser diodes emit optical energy over a narrow band of wavelengths, typically being only 3 to 5 nm wide (lasers are available that have very high purity output's, some being only 1 nm wide!). The chromatic dispersion loss figure conveniently nulls at 1310 nm, making this wavelength a good choice to use with standard optical cable. For interest's sake it is possible through the manufacturing process to shift the wavelength at which this zero crossing point occurs. For example, if longer distances needed to be covered and an operating wavelength of 1550 nm was desired, cable could be manufactured to have this null at 1550 nm instead of 1310. Cable is available that supports operation at multiple wavelengths simultaneously.

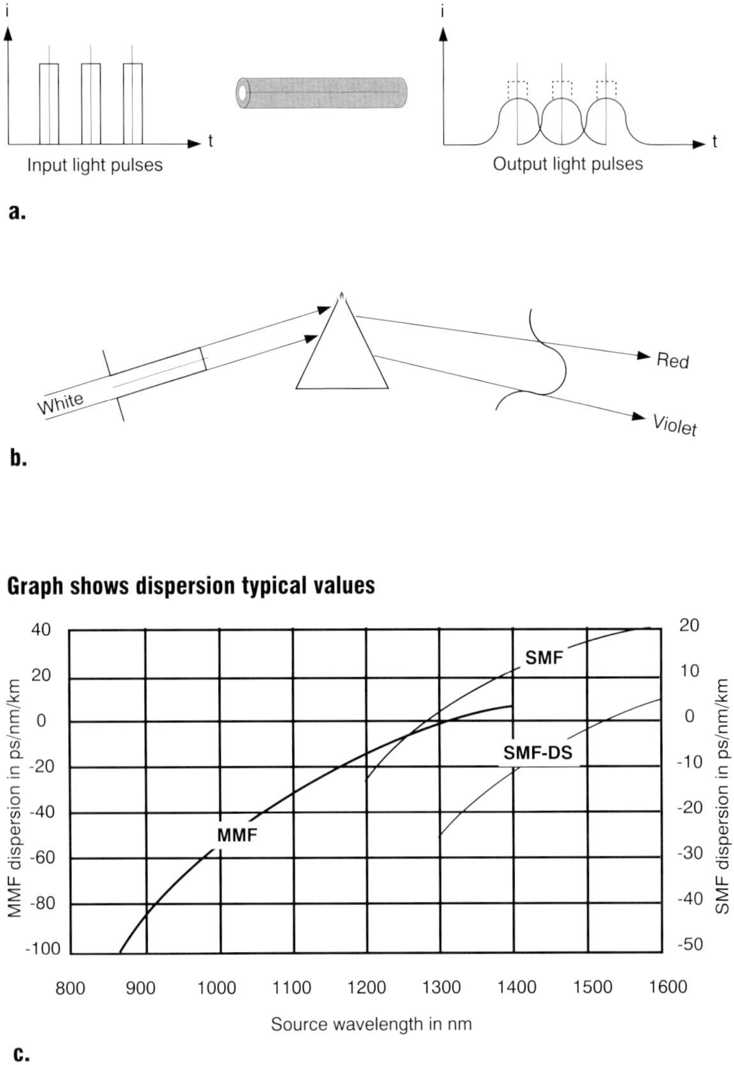

Figure 7-4. Modal and Chromatic Dispersion losses.

Other losses in the optical cable are well understood and easily predictable. They can be broken down into several broad categories.

Intrinsic losses are those attributed to the fiber itself. These include Rayleigh scattering, infrared absorption and contamination of the fiber during manufacture. Rayleigh scattering is caused by material variations and impurities within the optical cable. These impurities cause some of the light rays to be reflected such that they exit the cable (critical angle exceeded) or are returned to the source. The critical angle is the maximum angle of incidence that a light ray can have and still remain within the cable. Below this angle 100% internal refraction occurs, above this angle light escapes the cable and losses occur. Earlier we talked about the difference in indices of refraction between the core and cladding. These two materials determine the critical angle. For a glass fiber this angle is approximately 43 degrees. These same material variations and impurities also cause some of the light energy to be absorbed as it travels down the cable, attenuating the signal. Loss due to contamination is primarily due to hydroxal absorption. Water molecules within the fiber can absorb or reflect the light rays.

Extrinsic losses are usually caused by distortions in the fiber itself. These can include crushing type injuries caused by cable ties over-tightened during installation, as well as bends formed with too tight a radius. Following good installation practices can minimize these losses. The last major contributor is joint loss. Any time the fiber is spliced or a connector is installed there is loss. The amount of loss is determined by the type of connector used and the care exercised during its installation. Training and practice are the only ways to minimize these losses!

Reflections...Causes and Solutions

Reflections can be a source of loss within a transmission path and are typically created two ways.

The first is joint loss. When two pieces of fiber are joined, either as a splice or by using connectors, a butt joint is created (ignoring fusion splicing for now). Reflections are caused by errors in mating the ends of the fiber together correctly, and improper preparation of the fiber prior to installing either a connector or mechanical splice. At a junction between two pieces of fiber, light must be coupled out of one piece and into the other. The ends of both fibers must be parallel to each other and polished to remove scratches and other surface defects. Any misalignment, either due to the joints being offset or because the fibers are not touching, thus creating an air gap (dirt will create an air gap...see below), can cause reflections at the boundary. An air gap created by either of these problems will cause what are known as Fresnel reflections. Light is reflected when it encounters a boundary between two materials that have differing indices of refraction. Proper mating of the fibers will prevent or reduce these reflection losses.

The other is dirt. By and large dirt is responsible for more problems than any other issue, especially within a system that has just been installed. The width of the glass core in a single mode fiber is only 9 um, with cladding 125 um. For comparison a human hair is approximately 100 um in diameter. Contamination by even the smallest of particles will block a significant portion of the core area and cause a reflection to occur. During installation constant cleaning with isopropyl alcohol and lint free cloth is required to keep contamination to a minimum. Where connectors are used it is a good practice to clean them any time they are disconnected.

For the short runs found within a broadcast or production facility, finding the source of reflections can be challenging...unless you have the right tools. A Back Reflection Meter (BFR) and a simple attenuator made from a pencil will allow you to locate the source of reflections easily. In a properly working system back reflections will typically be 40 dB or more below operating level. An open fiber or disconnected connector (really one and the same) will show about –15 dB. Figure 5 presents a pair of NVISION

HD4270 optical transceivers used to provide a bi-directional HD video link between two floors in a facility. Assuming just one direction, there are four connectors present. Let us assume that after installation we show no signal at the receiver output on the ninth floor. A visual inspection of the installation reveals no obvious problems. We couple our BFR into the circuit near the transmitter and make our "pencil attenuator" by wrapping ten or so turns around a standard pencil. This is done to isolate just the length of fiber and connectors we wish to test. The reading we get is -50 dB...no problems here. We repeat the test at the connector junction in the first floor riser closet and also get a measurement of 50 dB. However, when we check the output of the ninth floor closet we get a value of –18 dB. Close examination of the connectors at the patch block shows there is a small air gap between the ends of the fiber. Reinstalling the connector fixes the problem. The HD4270s now provide a solid 1.5 Gbs connection between the two floors. In all likelihood this will be the last time you have to work on this gear !

Calculating Loss Budgets

At this point several applications have probably come to mind for which a fiber system can provide a simplified solution. As with any project, defining some up-front requirements will make designing and implementing a solution much easier. Applications that will be used within a typical broadcast or post production facility will generally not face problems regarding distance or bandwidth. More likely receiver overload will be a problem!

Causes of signal loss

Many of the losses encountered with fiber have been discussed earlier. For a practical transmission system these losses can be characterized as:

-Coupling loss

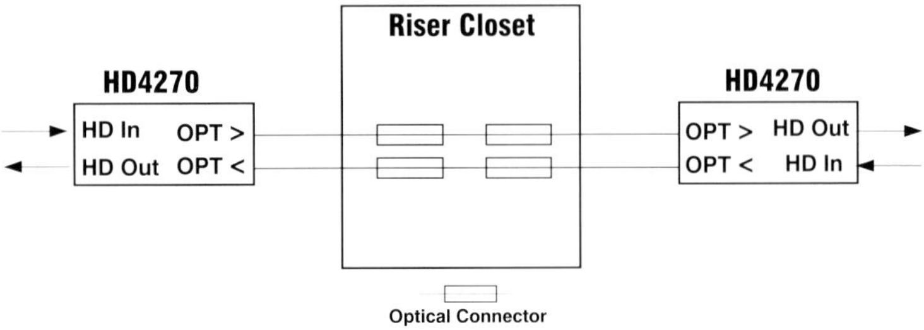

Figure 7-5. Interconnection between optical transceivers

-Optical fiber loss
-Connector loss
-Splice loss

Coupling losses are introduced at the transmitter output and receiver input. Optical coupling of the fiber to the laser diode or PINFET detector is not 100 % efficient. Factors such as the size of the core, the differences between the indices of refraction at the coupling, and the numerical aperture of the fiber will affect total coupling losses. NVISION equipment provides an optical connector already installed on the module interface card to aid in ease of systemization. Transmitter output power and receiver sensitivity is specified to include any coupling losses.

For loss calculation purposes these interconnect points may be treated as a connector. The optical transmitter typically has its output power rating specified as *launch power* and is rated in dBm. NVISION 4000 Series equipment provides a launch power rating that varies between -10 dBm and -3 dBm depending upon the specific module.

Optical fiber loss is simply the attenuation to the light energy as it passes through the fiber. The level of attenuation depends mostly upon wavelength used and fiber construction. If you are used to figuring loss using coaxial cable, fiber will surprise you! Typical loss values for a single mode fiber operating at 1310 nm are 0.35 dB per *kilometer*! 1300 nm is an important wavelength because attenuation and dispersion are low at this wavelength. Since the light emitted from the laser is at predominantly one wavelength, this attenuation does not vary appreciably with frequency, but instead is solely affected by distance.

In an ideal system there would only be two connectors present, one each at the transmitter and at the receiver. Real world installations may have patchfields, splices and other breaks in the fiber.

Connectors available today are fairly easy to install. When put on the fiber correctly, loss will be around 0.5 dB.

Splice loss can be very low, approaching 0.1 dB, if fusion splices are used. This process actually melts the ends of the fibers together. Unfortunately you probably won't have access to a fusion splicer, let alone the training necessary to use it correctly. More common mechanical splices that install much like a connector can be used instead. Figure about 0.5 dB of loss for each one. The best rule is to avoid splices entirely. Optical cable can be purchased in rolls up to 5 km or more in length, so splices really should not be required in a typical facility.

Rule of Thumb Calculations

With this information at hand, the maximum loss budget can be calculated. Loss budget is the difference between the transmitter output and minimum receiver input. A typical example:

TX output (-3 dBm) - RX minimum signal level (-15 dBm)
= loss budget (+12 dB)

This is the maximum allowable loss between the transmitter and receiver.

The transmitter output power or launch power is easily read from product specifications. The typical range is -3 to -10 dBm. Typical maximum power expressed in mW is 0.5 (-3 dBm = 0.5 mW). Receiver MDS (minimum discernable signal) is usually around -15 to -20 dBm. The input signal at which overload occurs is typically -8 dBm, giving a dynamic range figure of 7 to 13 dB. While this may seem restrictive, you will see below that it is more than adequate.

Next, the attenuation through the complete signal path needs to be figured. Assuming a distance of 1 km between transmitter and receiver, and single mode fiber operating at 1310 nm, the loss within the fiber would be a scant 0.35 dB! The run can be done with a single piece of fiber so there should only be two connectors, one at the transmitter and one at the receiver. These connectors are included in the product specification and do not count towards path losses. Our total path losses are just 0.35 dB *for a one kilometer run!!*

Subtracting path losses from the loss budget gives us +12 dBm - (0.35 dBm) = +11.65 dBm. This is the available "headroom" for this example. We could sustain almost 12 dB more loss before the path failed. This is another 33 km of cable! (11.65/0.35 = 33.1)

Looking at this from the perspective of the transmitter, we launch at -3 dBm. Our path losses are 0.35 dBm giving us a signal input at the receiver of -3.35 dBm. Our minimum signal level is -15 dBm so there is no problem here...lots of headroom. However, checking our specification for *maximum* input power we find it to be -8 dBm. We are severely overdriving the receiver and may be experiencing bit errors. Inserting an attenuator, or using the padded output port on the transmitter (many NVISION transmitters/transceivers provide this feature) to reduce the signal level will remedy this problem. It is possible, if the level is high enough, to permanently damage the photodiode in the receiver. SMPTE 292 specifies the point of electrical destruction as +1 dBm. Commonly available optical transmitters designed for video transport applications will not even begin to approach this level. Some Telco systems may drive near this power so exercise caution if interfacing to a Telco fiber.

A couple of typical fiber systems that might make sense in your plant are described below. These are just examples, so the actual numbers will vary with modules chosen and the path between source end destination.

Example 1

You currently have an NTSC facility with the transmitter site out on the back part of the property. Realizing the need to start transitioning to HD, you have just spent two years upgrading the plant internally to digital running at SD rates. Now the new transmitter and antenna are installed and you're faced with getting SD and HD signals and the associated audio out to the tower. There's not much left in the budget after all the other upgrades. Coax worked with the NTSC analog signals, but the SD/HD signals don't make it. You can't use equalization along the way because you'd have to put the rack of gear in the parking lot!

To simplify wiring, the optical transceivers will reside in your main equipment room near the front of the building. The physical distance the fiber will travel ends up being nearly 1000m or just over 3000 feet. There is no need to have a bi-directional path, so transceivers are not required. Instead an NVISION SD4171/72 has been selected to handle the SD signal, an HD4271/72 for the HD. Six channels of audio need to be sent so an SD4150/4155 at the station end embeds the AES audio into the SD data stream and an SD4160/4165 disembeds at the tower. The fiber pulled in is single mode, and the cable contained 10 individual strands allowing room for expansion if needed. Only two strands are required now.

The SD4172 has a launch power of -10 dBm; the HD4272 has a launch power of -3 dBm. At the tower the SD4171 minimum input level is -20 dBm for a loss budget of -10 dBm. The HD4271 minimum input power is -15 dBm providing a loss budget of -12 dBm. Both paths are identical, so fiber and connector losses total 2.4 dB for each (four connectors at 0.5 dB each and 0.4 dB for the fiber). Input levels to the receivers are -12.4 dBm and -5.4 dBm respectively. The SD4170 spec for maximum input power is -7.5 dBm so the calculated level will not require any adjustment. The HD4271 is rated at -8 dBm maximum input, so an optical pad will be needed to drop the level down somewhat. A 3 dB pad is needed, and looking at the cable specifications the manufacturer states that a two- inch loop will drop the level 0.5 dB. A 6 turn loop of fiber will do the trick. For safety this should be replaced with a proper pad, but for testing the loop will work fine. Alternately we could use the standard -8 dBm output on the HD4272.

Example 2

The rapid requirement for HD material has created a boom for your business, and you finally decide to put together a couple of full time HD edit suites. The problem is that the equipment is going to need a lot more real estate than you have available. A lease is secured on some empty space above you but it's three floors away. Initial testing shows that the HD signals won't go the distance. To complicate matters further, some of the gear you already have is first generation HD and cannot tolerate much input jitter. Several of the reclocking DAs you tried had enough jitter on some of the outputs that the older gear would not accept the signal. You also want to use as much of the existing facility as possible. An ENVOY HD router is purchased for the new addition. Tie lines to the old SD router allow the best use of your conversion equipment.

Since the tie lines are bi-directional, HD4270 transceivers are used. Six tie lines, three in each direction, are deemed sufficient so three modules are needed. Audio also needs to be sent, and although AES will go the distance it isn't worth the expense of pulling in extra audio cables when a single fiber optic cable with multiple fibers can do the job cheaper. Embedders and disembedders on the SD portion of the run will provide the audio link. Since the new router uses NVISION universal control, it interfaces easily with the existing router system. Another set of fiber transceivers designed for the computer network industry provides the data connection between the two router controllers.

The HD4270 transmitter has a launch power of -3 dBm, the receiver a minimum input level of -13 dBm. This provides a loss budget of 10 dB. There are several connectors and a patchfield located on each floor. The cable run is only 100 feet (about 30m) so cable losses can safely be ignored as they will be less than 0.1 dB. The combined loss through the connectors and such is 3 dB at each end for a total of 6 dB. This provides a receiver input level of -9 dBm, well above the minimum. The maximum level is -8 dBm so no additional corrections are needed. The other links for audio and data will calculate out similarly.

Practical Applications and Problems

Having lived with SDI for a decade or so, most of the systemization issues have been resolved and engineering an all-digital facility is not the challenge it once was. Most equipment on the market today allows virtual plug and play operation with SD data rates. The integration of HD into a facility has caused a resurgence of some of the issues that faced us years ago.

A pathological test signal is often used to verify link integrity since it stresses the equalization and PLL circuits within the receiver. Under normal conditions the average DC component present in an SDI bitstream will be zero. If long strings of ones or zeros are introduced this temporary DC component will drift away from zero towards a positive or negative value. This is not a DC shift in a real sense, but a low frequency AC superimposed upon the signal waveform. At the input gain stage of the receiver this DC component causes a shift in the operating point and can lead to non-linear operation. Additionally it can fool the equalization block, resulting in incorrect compensation further aggravating the problem. The result is an increase in bit errors.

In a typical fiber transmitter a laser diode is used to generate optical energy, which is modulated by the signal waveform. Refer to figure 7-6. Variations in intensity convey information to the receiver. As long as the modulating signal does not cause the variation in current through the laser diode to drop below threshold value, then very high levels of linearity can be maintained. An APC (automatic power control) circuit is used to keep the laser operating at a constant Q point, setting the steady state output power. The signal waveform amplitude modulates the laser out-

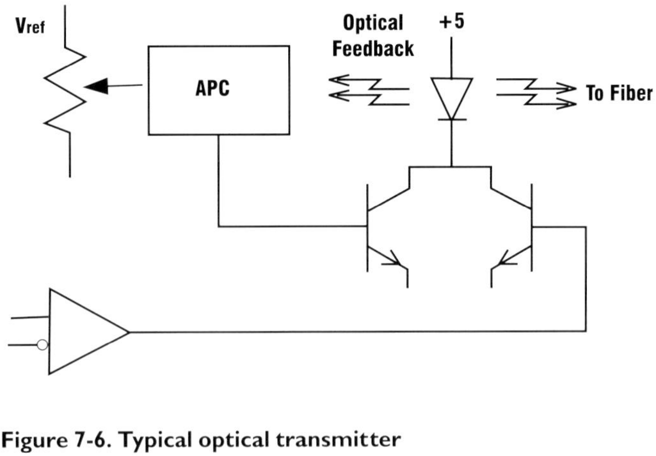

Figure 7-6. Typical optical transmitter

put. DC shifts present during long strings of similar bits can upset APC operation, shifting the Q point towards the laser cutoff or saturation regions, increasing the chances of bit rate errors being introduced into the data.

At the receiver (fig. 7-7), a PIN photodiode is used to convert the incoming light rays into electrical current. A trans-impedance amplifier simultaneously converts the current generated within the PINFET into a voltage signal and provides fixed gain to boost this voltage to a useable level. If excessive DC components are present on the incoming signal, this gain stage can be driven into non-linear operation. To combat this problem in typical designs, you would AC couple the signal between stages, removing the DC (really a low frequency AC signal) component. Unless care is taken to make sure the frequency response of the receiver extends low enough to pass this signal, bit errors will be introduced into the data. While this may reduce the severity of the problem, experience with real world video suggests that the bit patterns that induce these errors occur more often than originally predicted. Proprietary systems have been implemented to combat these problems, such as secondary scrambling of the bitstream, but by doing so compatibility is lost between various vendors' equipment.

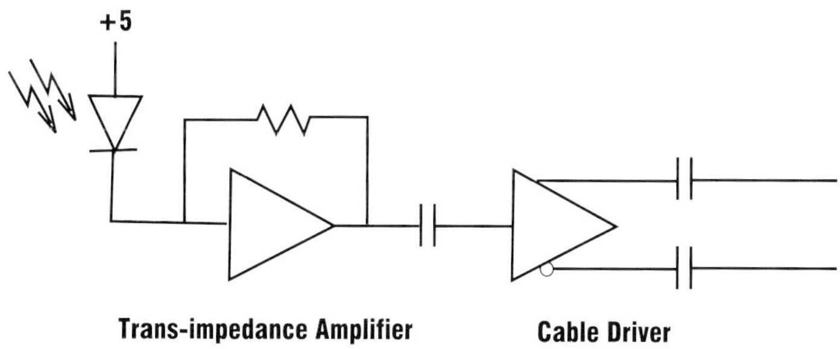

Figure 7-7. Typical optical receiver

These issues continue to be troublesome with SD and HD data rates. If left uncompensated severe bit errors can be introduced. Figure 7-8 shows an electrical/optical/electrical link using NVISION's 4000 Series transceivers (patent pending). The introduction of a VGA (variable gain amplifier) stage, and DC restoration compensates for the condi-

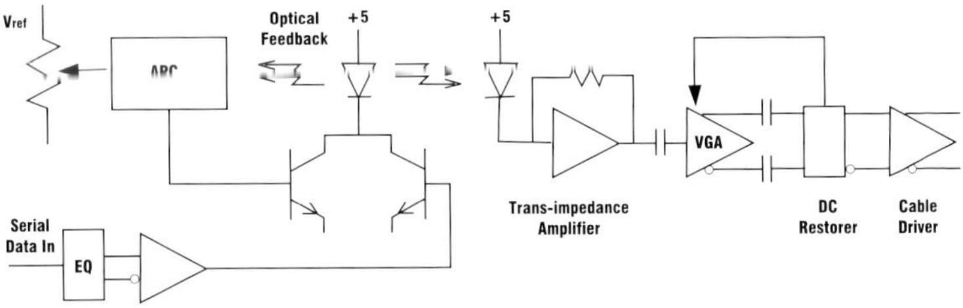

Figure 7-8. Block diagram of NVISION's Transmitter and Receiver

tion generated when pathological or other signals introduce DC components into the bitstream. A zero bit error rate is guaranteed no matter what the incoming bit stream looks like. The not so obvious benefit of this technology is that the standard NRZI signal is carried over the fiber allowing interface with many different vendors' transceivers. Using 4000 Series devices at both ends of the link provides much improved performance. See figure 7-9. When current equipment is upgraded or replaced with 4000 Series modules, compatibility is maintained with existing devices.

Troubleshooting

Locating and finding faults in a fiber system is not terribly different from what you are used to doing with traditional coaxial cable based installations. Attention to detail and physical examination of the installation will usually turn up the source of trouble. The issue for engineers and technicians dealing with fiber is usually a lack of practical

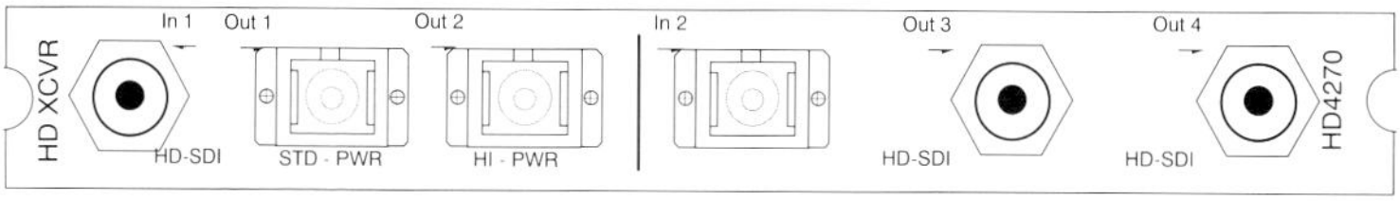

Figure 7-9. Typical transceiver connections. NVISION 4000Series

hands-on experience. If you are planning to integrate fiber into your plant some initial training would be paramount to making sure the endeavor is successful.

Some additional test equipment added to the shop will make troubleshooting much easier. As a minimum an optical power meter, an inspection scope and a back reflection meter (BRF) would be good investments. These tools will allow rapid fault isolation and easy preventative maintenance checks with a minimum of expense. If longer runs are used then an optical TDR should be obtained. A back reflection meter allows signal fault diagnosis on the shorter cable runs found within the plant that would be missed by a TDR. At the wavelengths involved, even the shortest pulse emitted by a TDR is still many meters long and cannot allow close inspection of losses present through connectors or on short cable runs. A back reflection meter coupled with an attenuator will allow easy troubleshooting of problems caused by excessive loss. A good supply of lint free lens cleaning cloth and alcohol will also come in handy...dirt is the enemy of fiber!

When a fiber system fails during commissioning the most common culprit is excessive reflections causing an increase in system loss that exceeds the available loss budget. Dirt is the primary culprit. As a result either no signal is seen at the transceiver output, or excessive bit error rates are seen. If the fiber runs point to point with no splices or patching then using a power meter to verify laser output at both ends of the cable will quickly show where the problem is. The inspection scope can be used to check the ends of the fiber for cleanliness and proper polishing. Typical problems to look for are air gaps between the ends of the fiber where connectors or mechanical splices are installed, fiber ends misaligned at fusion splices and the ever-present dirt!

Assuming no problems are found with the connectors or mechanical splices, the next step would be to dig out the back reflection meter to locate the source of the problem. Sharp bends in a single mode cable can allow light energy to escape the cladding and dramatically increase system loss. Wrapping several turns of the cable around a pencil can make a simple optical attenuator! Don't exceed the manufacturers minimum bend radius if you try this! Checking the cable run for inadvertent kinks and bends, especially near termination points will likely help in locating the problem. Visual inspection should be used first, but the BRF can be helpful in getting into the right general area.

Cable ties are another possible culprit for trouble. Any change in the diameter or structure of the cable can cause an increase in loss and/or reflections. Cable ties tightened down excessively can crush the fiber creating micro bends. These dramatically increase system loss. We do not recommend cable ties for fiber installation; instead use lacing twine or soft rubber straps. If ties must be used, provide a method of spreading the force out over more of the cable jacket. Keep in mind that anything that may knick or crush the cable should be looked for. Where the cable drapes over the end of a raceway or enters an equipment rack there could be sharp edges present that can cause a bend or kink to occur. Do not rely on the cable itself to provide mechanical strength, especially in areas where the cable must run vertically. If possible find some way to support the fiber and relieve any mechanical force that is being applied.

Chapter 8
Bringing It All Together

We have attempted to outline some of the various system components essential to building a broadcast or post production facility that incorporates the latest signal formats. As you read through this book, the specific facility you work in (or are responsible for) may or may not include many of the signal formats discussed. It is likely that at some point you will have to modify or upgrade your plant to add at least part of what we have presented. This final chapter will hopefully show you how to do this with the least interruption and the greatest chance for success.

Depending on where you reside, you are likely quite used to dealing with analog NTSC or PAL signals. A "typical" broadcast facility is shown in figure 8-1. The router forms the core of the plant, supporting many aspects of daily operation. Older facilities may only have the analog blocks; newer plants will likely have added some quantity of digital equipment using serial component or composite formats.

A well engineered component digital facility will find it relatively easy to add the additional hardware needed to support the HD signal formats. The ENVOY family of serial SDI/HD routers can be slaved to most popular control systems, making the addition of an HD layer to your current router painless. The requirement for multiple channels of audio can seem intimidating, with as many as eight chan-

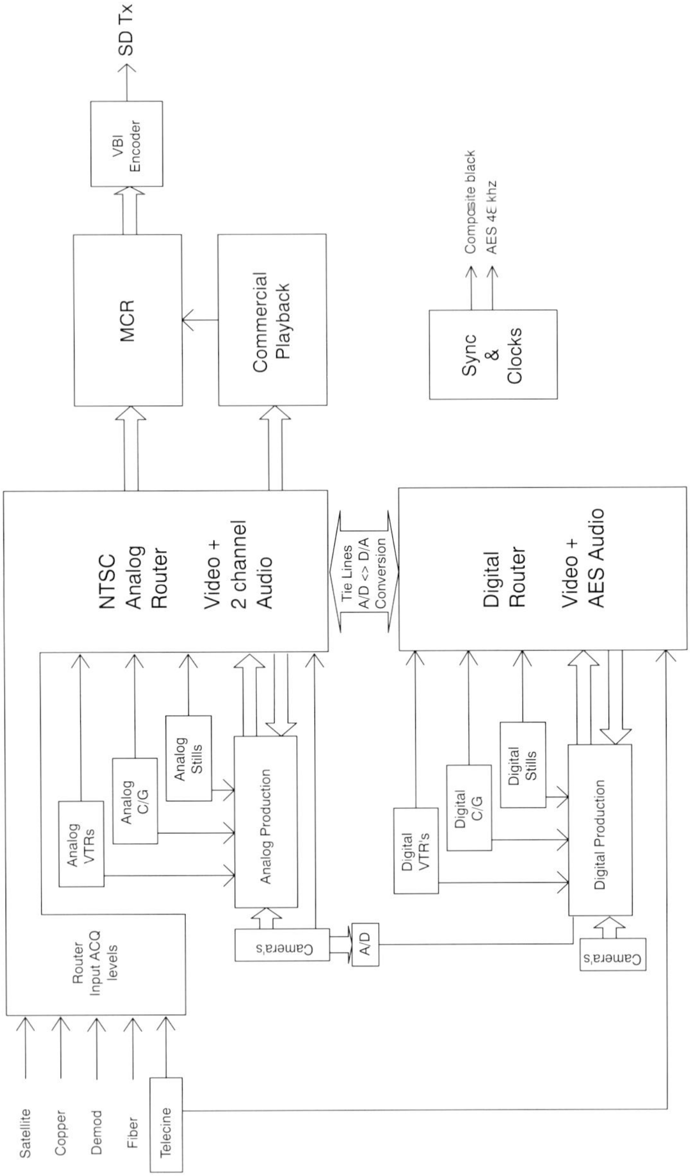

Figure 8-1. Today's Broadcast Facility

nels needed. NVISION NV3512 AES routers support large matrix sizes and multiple levels, simplifying this task. If your existing control system is not up to the task, NVISION's new ENVY router control system can be added on top to provide support for the additional layers that will need to be switched.

Paramount to success is making sure all equipment is properly locked to a common house reference. Figure 8-2 shows the timing relationship of the reference signals needed to achieve house synchronization. See chapter 15 in *The Book* for a further discussion of this and other concerns related to integrating AES audio and video.

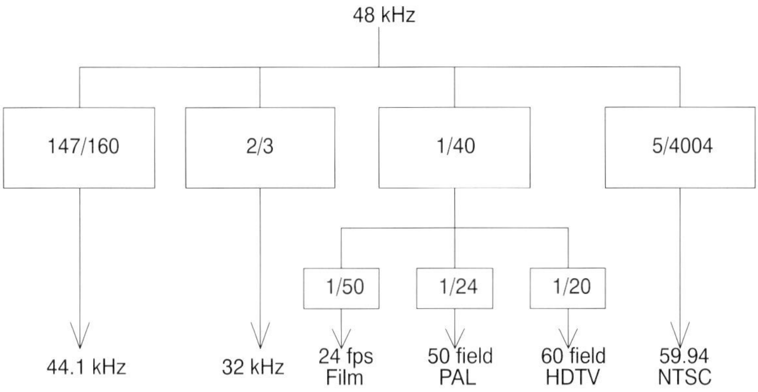

Figure 8-2. Multi-Format Synchronization Relationships

Some current HD equipment is capable of locking to house reference black. Devices not using color black will probably use a tri-level sync source as reference. Tri-level sync is unique in that the sync pulse runs positive and negative around blanking level. The synchronization point is defined as the zero crossing point on the positive going edge of the waveform. This is done to provide increased noise immunity and additional precision in determining the start of each line.

The easiest approach to creating this additional sync source is to slave a tri-level sync generator to your existing master sync generator. Keep in mind that this source may need to be significantly different in its timing relationship to house black, depending upon the equipment you are using. You are likely to encounter vertical timing errors. Current production switching equipment normally includes some form of auto-timing circuitry that will correct horizontal errors. Vertical errors pass through uncorrected, and if not accounted for can cause non-VI switching in routers, or picture offsets when passing through conversion gear. Auto-timing should not be considered a way to avoid proper reference system design; the input timing window is still finite in size and, unless care is taken to make sure sources are properly locked, difficult-to-diagnose problems can occur! Luckily, most of today's sync generators have features that allow independent timing of the individual outputs, as well as various test patterns to simplify timing adjustments. A sample timing chain is outlined in figure 8-3. A master oscillator running at 5 or 10Mhz is shown for those whose requirements demand high precision.

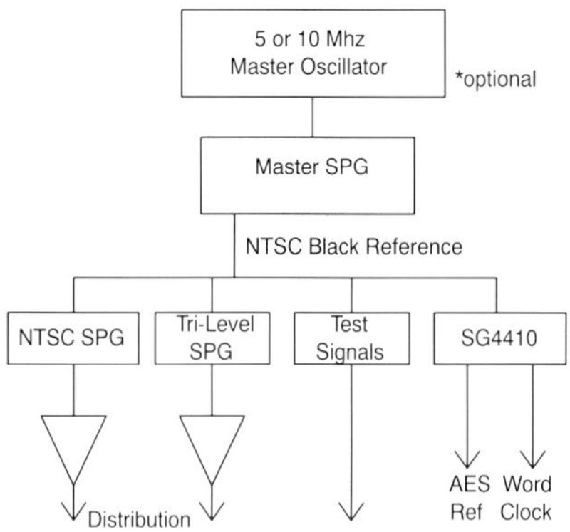

Figure 8-3. Representative Timing Chain

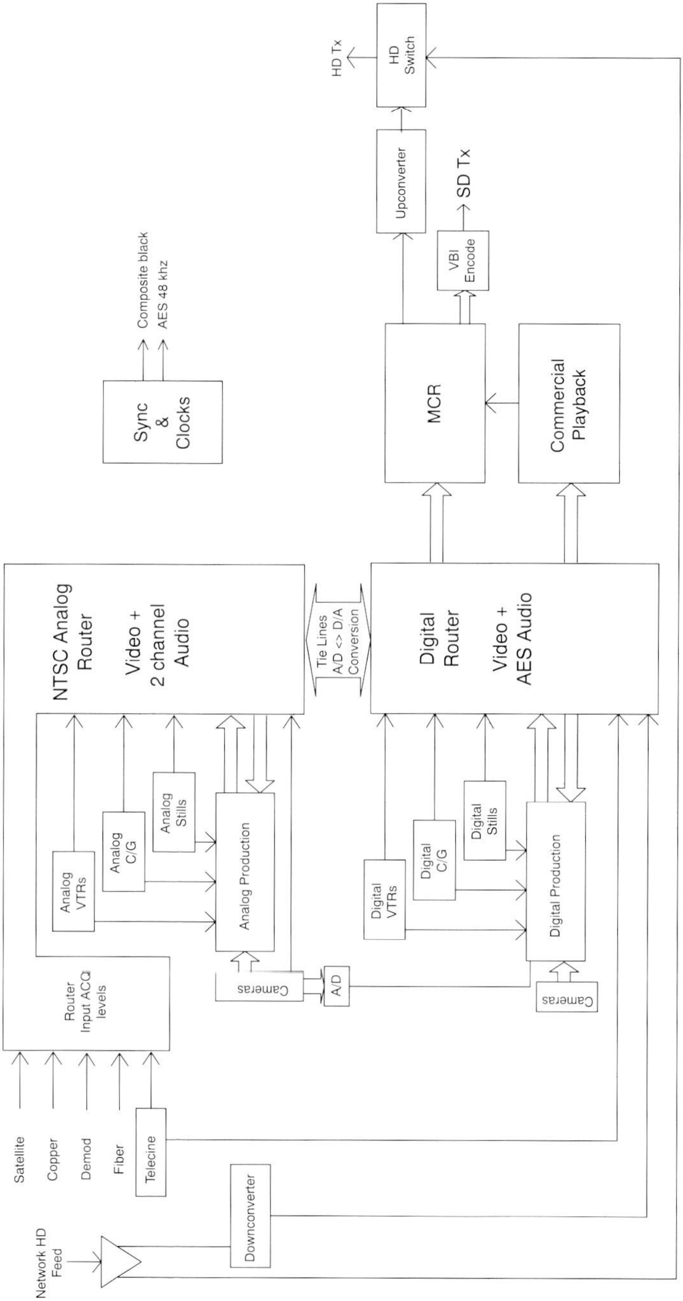

Figure 8-4. HD Integration into Broadcast

Integrating HD into your facility need not be difficult. Figure 8-4 shows what may be a common approach taken by many stations today. Notice the similarity to the typical facility outlined earlier. Only a couple of true HD sources will be present: program/commercial playback from HD VTRs and a network feed. All production is still done using SDI equipment which is up-converted to HD prior to transmission. Master control is left as SD to reduce the need for replacement of existing equipment. A down-converted version of the network feed is available to allow live hits to be inserted through production. More desirable is both HD and SD master control, slaved to a common control system. This way, as more HD equipment is brought online, less reliance on up-conversion is needed, and commercial integration is greatly simplified.

For on-air applications, synchronous audio switching is highly desirable, especially with the increased number of channels to be supported. As the quality of equipment available to the consumer continues to get better, more care must be paid to maintaining the integrity of the audio signals during production and transmission. Properly referencing audio gear as you do video equipment will ensure click-free and pop-free audio. Please see *The Book* for a thorough discussion of audio reference implementation and distribution.

The majority of HD material will originate from film sources initially, but this material will need to be converted to video for editing and distribution. A post environment will need to support a much larger range of formats than a broadcast house. Edit bays need to be configured on the fly to meet client demands. Auto-timing on production switcher inputs greatly simplifies timing issues, but care must be used to make sure that sources locked to tri-level sync and those locked to NTSC black are properly aligned for vertical phase.

Audio in a post environment is generally not hot-switched, so synchronous routing may not be required. NVISION AES routers are available in either sync or async configurations.

The ENVOY router family can support both SD and HD formats simultaneously within the same frame. They can serve as an input acquisition router ahead of the up-converters and also handle plant HD signals, whether they be local, network, or up-converted SD material. The up-converters are configured to reside within tie-line paths as well as on manually switched destinations. There will continue to be a need to deal with analog signals for many years. ENVY allows your existing router to remain useable throughout the transition. Tie-lines between your existing router and the ENVOY SDI level make efficient use of A/D conversion gear.

Audio processing can be handled by NVISION's 1000 and 4000 Series modules. A/D conversion using DA4030s into the NV3512 allows use of analog-only devices. NV1060 delay modules keep audio synchronized with the video. Additional audio processing to support multi-channel applications can also reside within these tie-line paths. One or more additional NV3512s support AES routing on the HD side. This provides support for as many channels of audio as you require. Both SD and HD master control allow multiple program streams to be output simultaneously. How these feeds are distributed is handled by an additional ENVOY router, configured for both SDI and HD. This capability provides maximum use of channel bandwidth, allowing single HD or multiple SD feeds to be routed to your transmission equipment as needed. Commercial integration with network or local HD material in either format is easily accomplished. Pay special attention to system installation. With bit rates of 1.5Gbits, quality cabling and connectors are required. Keep fiber in mind for long runs at HD rates. System timing should also be carefully

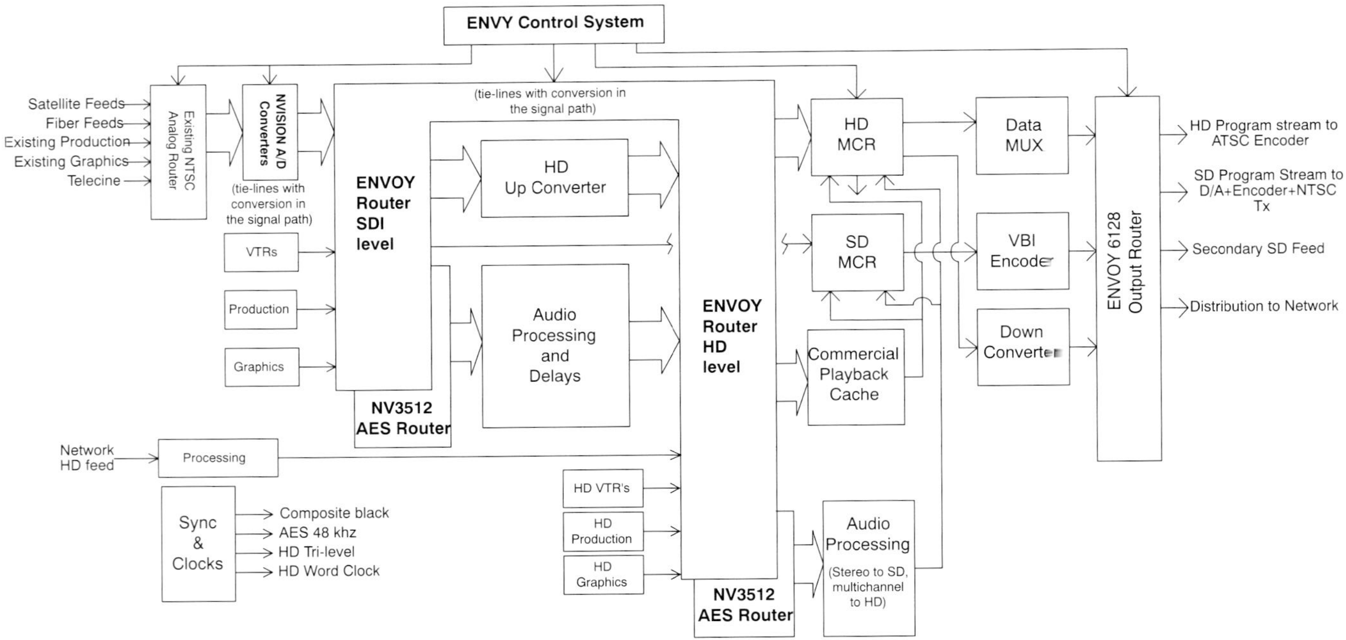

Figure 8-5. NVISION approach to HD Integration.

planned, as many different formats are present within the system and all need to co-exist with each other. Figure 8-5 shows a representative hybrid approach to a multi-format facility.

NVISION Products

Here is a brief outline of the products that NVISION currently manufactures. We may have added more since this book was published, so please contact us if you would like a complete and detailed catalog:

Modular Equipment

1000 Series—AES/EBU digital audio modules

2 RU equipment frame with redundant power option, for up to twelve modules
Distribution amplifiers
Analog to digital converters
Digital to analog converters
Digital delay compensators
Sample rate converters
Mix minus modules
Reference generators

4000 Series

1 RU frame, holds four modules
2 RU frame with redundant power option, holds eight modules

AES/EBU digital audio modules

Distribution amplifiers
Dual analog to digital converters
Dual digital to analog converters
Audio reference generators

SDI digital video modules
Distribution amplifiers
Fiber optic transmitters
Fiber optic receivers
Fiber optic transceivers
Audio embedders
Audio embedder input expansion
Audio disembedders
Audio disembedder output expansion

HD-SDI digital video modules
Distribution amplifiers
Fiber optic transmitters
Fiber optic receivers
Fiber optic transceivers

Routing Switches

Time Code routers from 8 x 32 to 512 x 512
Data routers from 64 ports to 256 ports
Asynchronous AES/EBU routers from 8 x 32 to 512 x 512
Synchronous AES/EBU routers from 8 x 32 to 2048 x 2048
SDI / HD-SDI routers from 8 x 8 to 256 x 256
SDI / HD-SDI secondary switches to allow expansions beyond 256 x 256

Routing Control Systems

Windows NT® Server, based control systems
Intelligent control panels
Control system GUIs
System APIs for user customization